Acacias
of
Australia
Volume One

Cover illustration: *Acacia podalyriifolia*

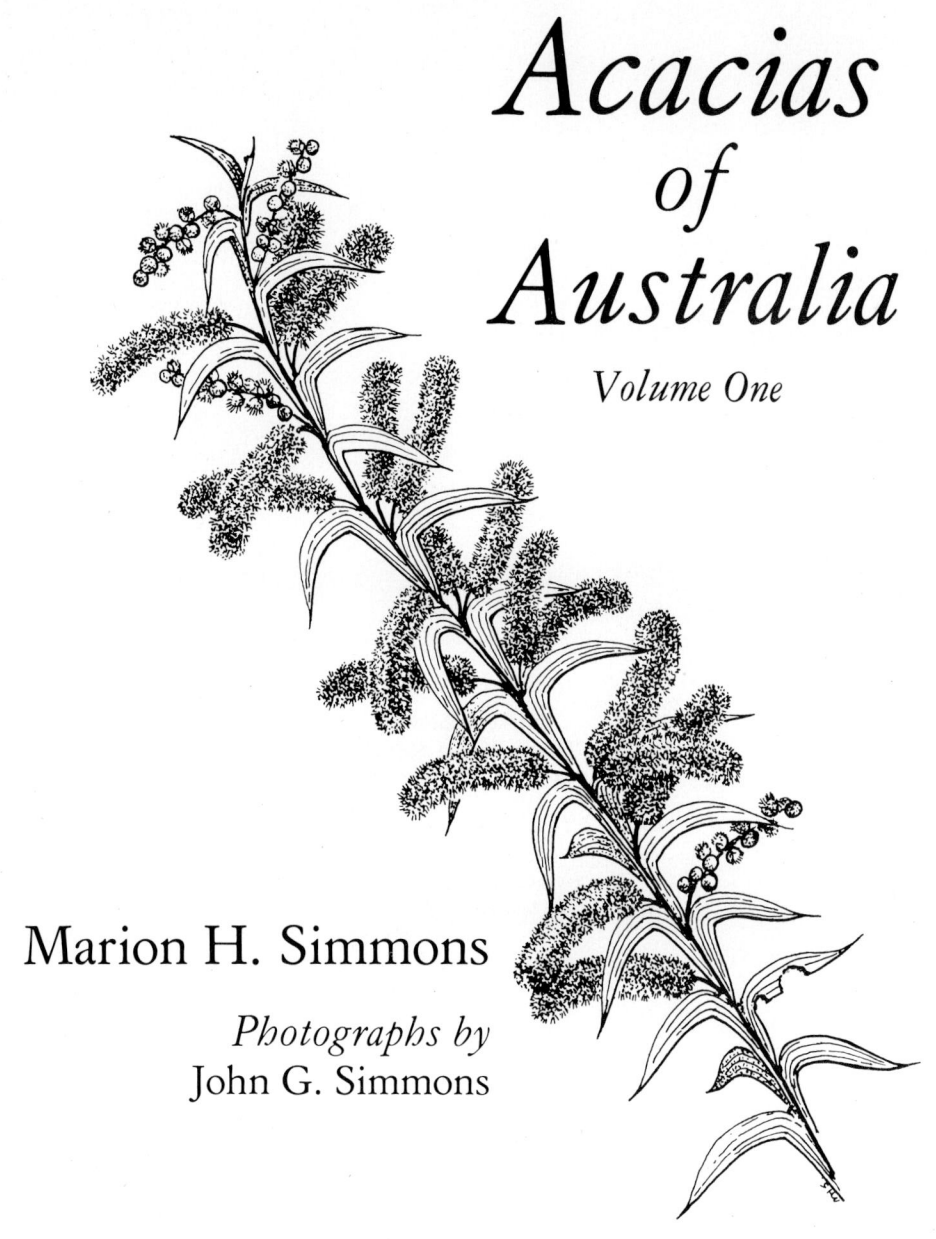

Acacias
of
Australia

Volume One

Marion H. Simmons

Photographs by
John G. Simmons

NELSON

Nelson Publishers
Thomas Nelson Australia
480 La Trobe Street Melbourne Victoria 3000

First published 1981
Reprinted 1982
This revised paperback edition 1987
Copyright © Marion H. Simmons 1981, 87

National Library of Australia
Cataloguing in Publication data:

Simmons, Marion H.
    Acacias of Australia. Volume 1.

    Rev. ed.
    Bibliography.
    Includes index.
    ISBN 0 17 007179 0.
    ISBN 0 17 007219 3 (set).

    1. Acacia — Australia. I. Simmons,
John G. II. Title.

583'.361'0994

Typeset and produced in Malaysia
by SRM Production Services Sdn Bhd

Foreword

Acacia in Australia is a large and complex genus considered difficult by both amateur and professional botanists. Though it has not received the attention that *Eucalyptus* has, in recent years there has been a considerable amount of taxonomic research on *Acacia*. Unfortunately, however, there is no comprehensive account of the whole genus or even of its members in Australia. There are a few keys and technical treatments of groups or of regions, usually written for the specialists. But until now the botanist, skilled or otherwise, has had no easy introduction to the genus, no overview of its complexity.

Acacias of Australia is no dry tome, the result of a study of dried plants carried out in a musty room somewhere in the city, but a living book on an intriguing and important group of plants. Technical details have not been neglected, but a book such as this could not have been produced without hours in the bush. It is the result of hard work and careful planning: extensive travel over miles of rough roads, sketching and photographing living plants, and many hours of discussion with people who know *Acacia* in the field, in the garden, or merely as lifeless scraps for scientific study. Marion Simmons's book, with its descriptions and beautiful illustrations of so many species from so many parts of the continent, is an admirable introduction to *Acacia* in Australia.

LES PEDLEY
Assistant Director
Queensland Herbarium

Acknowledgments

Many people have given encouragement and assistance during the preparation of this book. I would like to acknowledge and sincerely thank them all for their generous help.

My husband, John, first of all for his photographs and because without him there would be no book; Gwenda and Ross Macdonald for their continued encouragement, generous hospitality and practical help in so many ways; an especial thank you to Dorothy and David Gordon, Val and Roy Hando of Queensland, Ron Smith of Tasmania, George Althofer of New South Wales, Kate and Bruce Copley of South Australia; my thanks also to Merv Hodge, John Donohue, Kerry Davis, John Hamilton of Queensland; Dr Neville Marchant, Ken Newbey, Gary Phillips, Ray Reed, Eileen and Cliff Croxford, Thelma and Pat Daniells of Western Australia; Audrey and Athol Beswick, Hazel and Ken Dean of Tasmania; Angus Torpey, Trevor Blake, Bill Perry, Diana and Fred Bienvenu of Victoria.

I would also like to express my appreciation of the cooperation and invaluable assistance which has been given freely by the *Acacia* botanists; in particular to Les Pedley, Assistant Director of the Queensland Herbarium, for advice given on collecting areas, for the identification of many specimens, for checking descriptions and for writing the introduction to this book; to Dr Mary D. Tindale, Principal Research Scientist of National Herbarium of New South Wales, Bruce Maslin of the Western Australian Herbarium, David Whibley of the South Australian Herbarium, J. R. Maconochie of the Arid Zone Research Institute, Northern Territory, for identification of many specimens and for checking descriptions; to Mrs Mary Cameron, honorary botanist, Queen Victoria Museum, Launceston, for checking descriptions and to Bernie Hyland of the Division of Forest Research, CSIRO, Atherton, Queensland, for assistance given and for permission to work in the herbarium. My thanks also to the Curator of the Western Australian Herbarium for permission to work in the herbarium and to the staff for their assistance; to the Forestry departments of the various states for the issuing of collecting permits; especially to the Forestry Department staff at Baradine, New South Wales and Kalgoorlie, Western Australia for maps, information and assistance given during our travels.

Contents

List of Colour Plates

Introduction

The 'acacia project', as we were soon to call it, began seven years ago and evolved from a simple decision to put together all available information on the Australian genus *Acacia* (the wattles), which was scattered through a great number of botanical and gardening books. We were finding it difficult, sometimes impossible, to locate any information at all on some of the plants we were growing.

From there the project gradually took shape. Collecting trips became a highlight of our existence; photography began in earnest; descriptive details were researched; specimens were collected and pressed; the illustrations began, at first as pencil drawings, later as the line drawings which have been used to illustrate this book.

In the early stages we found that Ron Smith of Hobart also was gathering together early descriptive botanical material. When he learned of our project he generously gave all this material for our use, and for some time continued to help with the collection of copies of the early records. This assistance added great impetus to our own efforts.

Many of these early records had been published in Latin for which no English translations were available. Our good friends, Gwenda and Ross Macdonald of Croydon, Victoria, came to our rescue on many occasions and provided us with a long list of translations.

During the eight years since the project began, we have utilised school holidays and long service leave to travel as widely as possible in some of the more remote areas of Australia collecting, photographing and recording acacias for inclusion in this book. A standard *Ampol Touring Atlas* has been used as our basic road map. Where necessary this has been reinforced with more detailed local instructions. Frequently tourist maps giving navigable tracks which are not shown on any other map are available at different towns.

Some official maps for the more remote areas are not always accurate and should be used with great caution. Tracks unsuited to the family car are sometimes shown as normal roads on these maps, others which are shown do not exist at all, and vice versa. Checking with authorities at the last township is essential before travelling on these lonely tracks.

As yet we have not visited more northern regions, especially north of Carnarvon in Western Australia, the Northern Territory and far northern Queensland. Where material from these areas has been used for drawings, it has come from collections made by friends travelling through, and subsequently has been identified by botanists of the appropriate herbarium.

One hundred and fifty species have been included in the book, a number from each state. It will be realised that only a few species occur in nearly all states, many in several and a relatively few are restricted to only one. Western Australia has the highest number of endemic species.

Illustrations have been drawn from material – some fresh, some dried – which has been collected during our many trips and which has been identified by botanists of the various state herbaria. Colour slides have proved an invaluable backup for the drawings.

1

Each species illustrated by a line drawing has been drawn to natural size, showing part of the branchlet, flowers, pods and some seed details. Habit and other details of some species have been illustrated by colour plates.

Often there is great variation in size and sometimes shape of the phyllodes of the one species, and where material has been available this has been shown in the drawings. It must be appreciated that the illustrations cannot represent all the variations, however the accompanying descriptions do outline the extremes as they are known. Plants are not always well collected or known from some of the more remote areas of Australia.

All species have been described as fully as possible with a minimum use of technical botanical terms. Any terms needing explanation have been included in the glossary.

Common names have been given where they are known. At times the use of common names can be confusing because more than one common name may be used for the same plant in neighbouring states, or the same name may be used for two or more species. For instance, Prickly Moses in Victoria and Tasmania is *Acacia verticillata*; in New South Wales it is *Acacia ulicifolia*; and in Western Australia it is *Acacia pulchella.* Australian birds have been given official common names, but to date plants have not. Until this occurs it is preferable to know and use the correct botanical name.

The common name 'wattle' was used in the term 'wattle and daub' and referred to the combination of interwoven saplings and mud which was utilised by the early settlers in Australia to construct their shelters. The most generally used material was not a wattle but *Callicoma*, an entirely different shrub with wattle-like flowers which was commonly called Black Wattle. The name wattle eventually became associated with the acacias.

Name changes and changes of status may occur before publication, since botanical review is being carried on continually by botanists of the state herbaria. To ensure accuracy all descriptions have been checked by state botanists interested in the study of acacias. The most recent synonyms have been included for the sake of clarity.

Plants with similar characteristics have been grouped together for ease of identification and comparison. A simple key to the groups has been included on page 14. The sizes of the plants have been given as they are known from field observations and from previously published botanical data. It must be stressed that these sizes should be taken as a guide only, because plants in cultivation often attain a far greater size when they are grown in ideal garden conditions. Plant dimensions given in the text are height or length × width, in this order.

Flowering times, too, may vary with different climatic conditions. Those given are the usual flowering times, but quite often plants growing in Tasmania, for instance, may flower some time later than the same plants growing in Queensland. This is not always a consistent difference. Altitude also may influence flowering times.

For those who are interested in growing some of our beautiful acacias, whether for their foliage, flowers, perfume or for more practical uses, there is a wide variety from which to choose, from tiny creeping prostrate shrubs to towering forest giants. There are many smaller plants which are well suited to growing in suburban gardens, as container plants or in rockeries, and others which are suitable for use as shade trees, in windbreaks, roadside plantings, parklands and in farm forests. Many are tolerant of a wide range of climatic and soil conditions, while others, especially from northern Australia, sometimes are very slow to grow in the colder southern states and may flower but not set seed; on the other hand, some of the well-known southern acacias

2

will grow in the north but are reluctant to flower or, if they do, flower only sparsely. Careful choice of species will avoid these problems and will make it possible to have acacias flowering in your garden at any time of the year.

Acacias are often said to be short lived and with a life span of about five years for *Acacia pulchella* and ten to fifteen years for *Acacia baileyana* and *A. podalyriifolia* this may seem to be true. However, it will be appreciated that unfavourable conditions of climate or situation may shorten a plant's life. Attack by insects, such as wood borers, also is considered a factor in a plant's early death. In spite of this, there are many species which are known to live for many years and some to a great age, like *A. melanoxylon* (Blackwood), *A. pendula* (Myall), *A. aneura* (Mulga), *A. excelsa* (Ironwood), *A. papyrocarpa* (Western Myall) and the remarkable *A. peuce* (Waddywood) from the inland deserts.

Society for Growing Australian Plants

This society, of which we have been members for many years, has been instrumental in the collation and dissemination of a wide range of information on native plants in Australia through their quarterly journal *Australian Plants*, and other publications which appear irregularly. There are branches in all states for people who are interested in growing native plants. Each state branch or regional group meets monthly, issues a quarterly newsletter, offers a free seed bank service to members and conducts outings to places of local interest.

Within the society, study groups are formed to study different plant families of our unique flora, to enable a wider knowledge, appreciation and experience to be gained and shared with others. Among these study groups there is one concentrating on acacias. Regular newsletters and an extensive seed bank are available to those who become members. Membership of the society is a pre-requisite to membership of any study group.

Introduction to the Second Edition

Publication of recent floras, the results of botanical research and the collection of many more specimens have expanded the information available on many of the acacias included in the first volume.

Specific details have been corrected where necessary in the second edition and several descriptions have been altered where research has indicated that a division of species was warranted. This applies in particular to *A. acinacea*, *A. buxifolia*, *A. calamifolia*, *A. chrysella* and *A. microbotrya*.

Some uncertainty exists at present regarding the status of the name *Acacia*. A paper by Les Pedley published in the Botanical Journal of the Linnean Society, London, in 1986 divides *Acacia* into three smaller groups. If this were generally accepted almost all Australian species would be transferred from *Acacia* to *Racosperma*.

The publication *Draft Index of Author Abbreviations* compiled at The Herbarium, Royal Botanic Gardens, Kew, England, in 1980 has changed the way in which plant authors' names should be written. For instance, A. Cunn. should now be written as Cunn., Tindale as Tind. and Lindl. as Lindley. However, this style has not been adopted for the second edition as the system used in the first edition also clearly denotes the plant authors' identities and is still in common usage. See pages 317 and 318 for an explanation of abbreviations of plant authors' names used in this book.

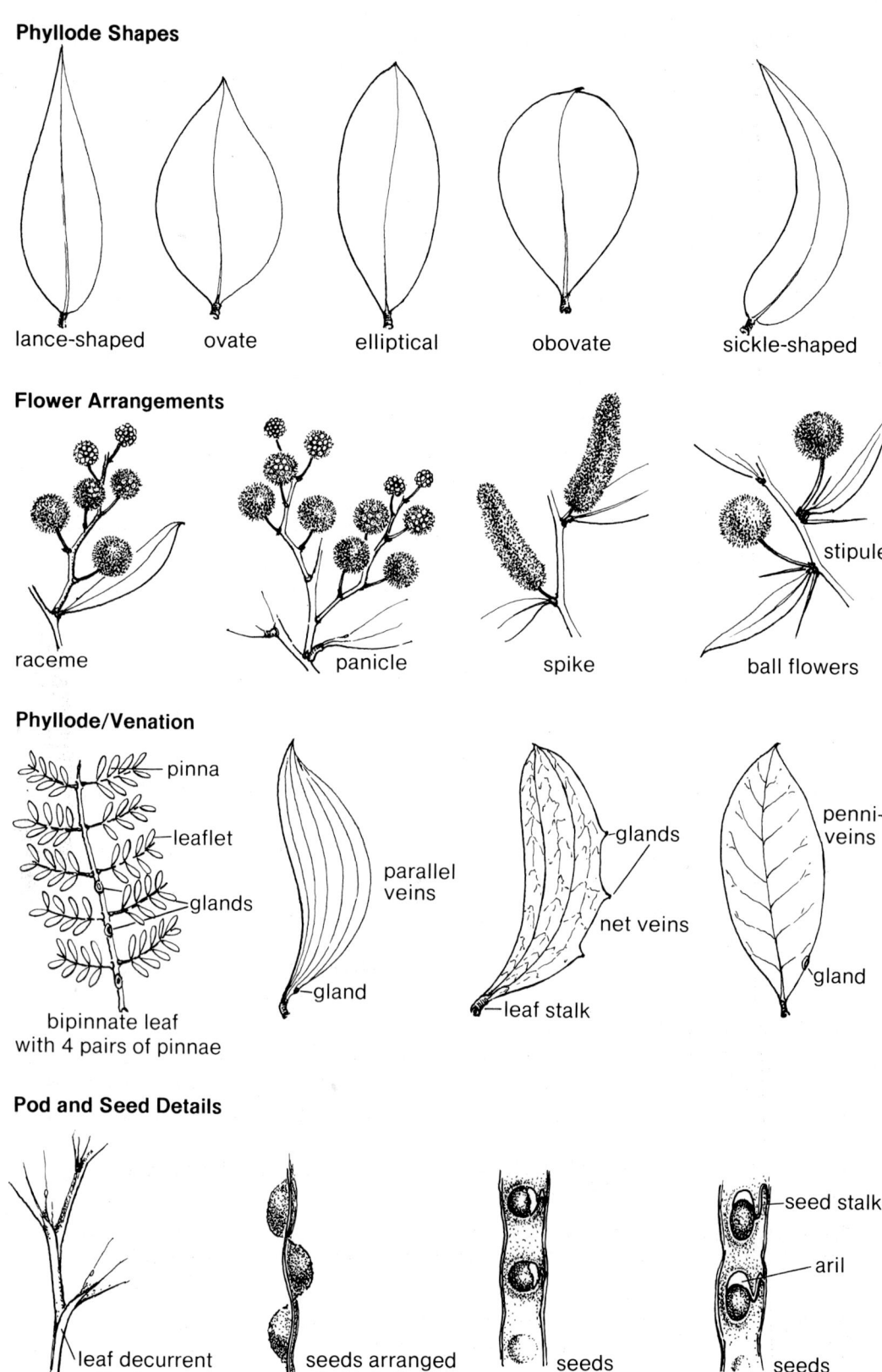

Phyllode Shapes

lance-shaped ovate elliptical obovate sickle-shaped

Flower Arrangements

raceme panicle spike ball flowers

stipules

Phyllode/Venation

pinna
leaflet
glands

bipinnate leaf
with 4 pairs of pinnae

parallel
veins

gland

glands
net veins
leaf stalk

penni-
veins

gland

Pod and Seed Details

leaf decurrent
with stem

seeds arranged
alternately

seeds
transverse

seed stalk
aril
seeds
longitudinal

Glossary

alternate	arranged singly in two rows, but not opposite
appressed	closely pressed against surface
aril	swollen portion of stalk attaching seed to pod, often covering part of seed
axil	angle between phyllode and stem; thus axillary
bipinnate	twice divided compound leaf
bract	small leaf-like structure at base of flower-stalk or raceme
bracteole	small bract on flower-stalk directly below flower
calyx	outer ring of a flower; the sepals as a group
corolla	inner ring of a flower; the petals as a group
deciduous	falling early, referring to bracts or phyllodes
decurrent	running down; where base of phyllode continues down stem as a raised line or ridge
gland	a projection or insertion on phyllode margin, nerve, or on stalks of bipinnate leaves; a nectary
glaucous	covered with whitish bloom
hoary	greyish, mealy surface, caused by dense covering of very short hairs
leaflets	smallest leaves of a bipinnate leaf
linear	long, narrow, straight-sided
lobes	partial division of any part
minni ritchi	referring to bark of an acacia which peels off in thin or tightly curled pieces
nerves	or veins are threads of connecting tissue branching within the phyllode or leaf
node	point of stem where phyllodes arise
oblique	unequal, off-centre
oblong	longer than broad, more or less straight sides
panicle	flowering branch comprising several racemes
penniveins	pinnate or feather-like veins
petal	one part of inner whorl of a flower, usually coloured
phyllode	flattened and enlarged leaf stalk which looks like and functions as a leaf
pinna	first division of a compound leaf consisting of a stalk and leaflets; plural: pinnae
pinnate leaf	a compound leaf with leaflets on either side of a common stalk
pubescent	downy, softly hairy
pungent	sharp-pointed
raceme	several stalked flowers arranged on a common stalk, lower flowers opening first
recurved	bent or curved backwards
reflexed	bent sharply backwards or downwards
resinous	sticky

scabrous	rough to touch
sclerophyll	hard leafed; often describing eucalypt forests
scurfy	scaly, rough
sepal	one part of outer whorl of a flower
sessile	without a stalk
shrub	woody plant divided into separate branches from or near ground level
spathulate	broadest at tip; spoon-shaped
species	a member of a group of plants with similar characteristics
spike	a rod of stalkless tiny flowers, the lower ones opening first
spine	stiff, sharply pointed appendages
stipules	two projections, soft or stiff, at base of phyllodes
synonym	a plant name which is no longer in use
transverse	cross-wise
undulate	wavy
viscid	sticky
whorl	referring to a ring of phyllodes or other parts arranged evenly around a stem

The Genus Acacia

Even a casual observer would be aware that the vegetation of Australia is dominated by *Eucalyptus* and *Acacia*. There are few places outside of the most urban of urban areas that a plant of one or other cannot be seen. *Eucalyptus* is virtually confined to the Australian region, but *Acacia* is usually described as pantropical. Not very accurately. It extends to the subtropics in the northern hemisphere, not occurring at all in Europe, but it extends beyond the subtropics in the southern hemisphere. Note the Tasmanian endemic, *Acacia pataczekii* (see colour plate 1). No one really knows, but reputable scientists have estimated that there are more than 1,200 species of *Acacia*, more than 700 of them native to Australia.

In recent classification *Acacia* has been regarded as consisting of three subgenera. One has bipinnate leaves with spines in their axils. The flat-topped trees associated with the giraffes and elephants in the African landscape are *Acacia* of this subgenus but there are also species in Asia, South America, and perhaps a dozen in northern Australia. *Acacia farnesiana* (see colour plate 2) is typical of the subgenus as a whole but not a typical Australian representative. Although it is widespread in Australia and was here prior to European settlement, it appears to have come originally from Central America. The second subgenus consists mainly of prickly woody vines or small trees widely spread in the tropics but with only one Australian species. This is *Acacia albizioides*, a woody vine found in rainforests of Cape York Peninsula. The rest of the Australian species belong to the third subgenus, members of which usually have no thorns or prickles and leaves modified to phyllodes. Though Western Australian species such as *Acacia drummondii* (see colour plate 3) and *Acacia varia*, and eastern Australian ones such as *Acacia decurrens* and *Acacia terminalis* (see colour plate 14) have true leaves, in other characters they are related to species with phyllodes and belong to the 'Australian' group. A few species of the group are found outside the Australian region as far afield as Mauritius, Formosa and the Hawaiian Islands. All species of *Acacia*, whether they ultimately produce phyllodes or not, have true leaves in the seedling stage. Some phyllodinous ones such as *Acacia rubida* may be a few metres tall and produce flowers before phyllodes are developed, while others may produce only two leaves before the first phyllode develops.

Species of *Acacia* are found virtually throughout the continent though only a few are found in rainforest. The south-western part of Western Australia and to a lesser extent south-eastern Australia are rich in species while, except in rugged sandstone areas, northern Australia has fewer species. Large areas of the inland are dominated by pure stands of single species such as Mulga, Brigalow and Gidgee.

The range of variation in *Acacia* in Australia is so large that at first it might be thought desirable to split it into different genera. Compare for example *Acacia bancroftii* with *Acacia cometes* (see colour plates 4 and 16). There is, however, uniformity in the structure of the flowers. Except in degree of division of the calyx, or degree of hairiness, or ratio of corolla to calyx, all *Acacia* flowers look very much the same. The arrangement of flowers in the inflorescence does vary but the variation is not great. The range of flower colour is also not great. Shades of yellow, from almost white to orange-yellow are the rule. The outstanding exception is *Acacia purpureapetala* (see colour plate 10).

It is in the phyllodes and habit of growth that the tremendous differences between some species are most striking. The average length of the phyllodes of *Acacia minutifolia* is probably about 1·5 mm while those of *Acacia dunnii*, a tropical species, are commonly more than 30 cm long. Phyllodes are usually regarded as an adaptation to an arid climate, but in Australia the phyllode has developed in all sorts of ways regardless of climate. Differences in habit and size of plants are equally striking. *Acacia depressa* is a compact, prostrate, cushion-like shrub 2 to 5 cm high while *Acacia bakeri* is a rainforest tree up to 35 m high.

Despite the great diversity of *Acacia* in Australia and in other parts of the world, the genus has some distinctive characters not found in Australia. In central America and in Africa there is a remarkable association between some species of *Acacia* and ants. Both benefit from the association. The ants feed on nectar, which is rich in nutrients, secreted from nectaries ('glands') on the axes of the leaves and for their part, the ants discourage other insects from eating the foliage of the *Acacia*. The plant even provides living quarters for the ants in hollow spines which are much larger than those found on other species of *Acacia*. There are no Australian 'swollen-thorn' acacias which shelter ants but most species do have glands on the leaves. There may be some association between ants and *Acacia* but it is certainly not a well-marked one. The glands of some species secrete nectar and ants may be attracted to this but the glands of many species do not appear to function at all. They may do so for a short period in the seedling stage of the plant, attracting ants when protection from predators is most needed.

Whatever their function, glands are useful in identification. They vary in structure and position from species to species. They are found usually on the rachis of bipinnate leaves and on the edges of phyllodes. *Acacia leptospermoides* and *Acacia spathulifolia* are representative of a group of Western Australian species which are unusual in having glands on the surfaces of the phyllodes and *Acacia bancroftii* and *Acacia wardellii* have glands on projections from the phyllode margin.

Most readers of this book would agree that acacias are interesting and often attractive plants but from their published journals it seems that early European visitors to Australia were not particularly impressed by them. In fact they found the vegetation as a whole rather unattractive. Joseph Banks described the landscape south of Sydney as seen from the *Endeavour*: '. . . it resembled in my imagination the back of a lean Cow, covered in general with long hair, but nevertheless where her scraggy hip bones have stuck out farther than they ought accidental rubbs and knocks have intirely bared them of their share of covering.' Unflattering descriptions of this sort were common until about the middle of last century. It took naturalists as long to appreciate the harsh beauty of this country as it did the poets and painters.

Although they may not have written about the country in glowing terms, these visitors recognised that Australian plants were different. They collected them and took specimens back to Europe. Carolus Linnaeus had developed a system of classification of plants using specimens from all continents except Australia. These strange plants collected in New Holland were readily accommodated in the system and before long the Linnaean methods were generally adopted by botanists. Soon after the return of Cook's expedition to Britain plants from Australia were being described in scientific publications.

1 *Acacia pataczekii*

2 *Acacia farnesiana*

3 *Acacia drummondii*

4 *Acacia bancroftii*

5 'Minni-ritchi' bark of *Acacia* species

6 Habit of *Acacia grasbyi*

7 *Acacia myrtifolia*. Phyllode with one longitudinal nerve

8 *Acacia alata* var. *biglandulosa*. Phyllodes winged and decurrent with stem

9 Pods of *Acacia oswaldii*

10 *Acacia purpureapetala*

11 Pods of *Acacia leichhardtii*

13 Pods of *Acacia complanata*

14 *Acacia terminalis.* Bipinnate leaves

12 Pods of *Acacia terminalis*

15 *Acacia ensifolia*

16 *Acacia cometes*

17 Foliage of *Acacia phaeocalyx*

18 *Acacia doratoxylon.* Flowers in spikes

19 *Acacia calamifolia*

20 Foliage of *Acacia gladiiformis*

21 *Acacia platycarpa.* Flowers in globular heads; phyllodes with several longitudinal nerves

22 *Acacia leptospermoides*

23 Flowers and pods of *Acacia meisneri*

24 *Acacia flexifolia*

25 Trunk of
Acacia pentadenia

26 *Acacia pentadenia*

27 *Acacia juncifolia*

28 Foliage of *Acacia uncifera*

29 *Acacia harpophylla*. Brigalow

30 Pods of *Acacia implexa*

31 *Acacia doratoxylon*. Woodland

One of the first Australian plants collected by a European was *Acacia truncata*, which is common in coastal heath in the south-western part of Western Australia. It was described and illustrated by a Dutch botanist, N. L. Burman, in 1768 as *Adiantum* – a fern! Burman may be excused for his error as the specimen he saw had no flowers and pods and its phyllodes are not unlike the leaflets of a fern. Burman believed his specimen had come from Java but it is likely to have been collected by Willem Vlaming near the Swan River in 1697. William Dampier had collected specimens of two species of *Acacia* on his two voyages to Western Australia but these were not critically examined until Australian plants had become widely known.

Not only dried specimens but seeds were taken to Europe and more were collected soon after the settlement of Port Jackson was established. By the early part of the nineteenth century acacias were growing, often in glasshouses, in France, Italy, Austria and Germany as well as in Britain. Some species thrived and have in fact become naturalised in southern Europe. Descriptions of these cultivated plants were published in gardening literature. Some of the descriptions were supplemented with good illustrations and the plants can be accurately identified, but other descriptions were inadequate and some of the plants described may never be identified with certainty. By 1825 about one hundred species had been named, about a third of them from plants in botanical gardens. French and British expeditions had collected Western Australian plants chiefly near the coast but after settlement in 1829 collectors began sampling the rich flora of the south-west and herbarium botanists worked on their specimens. Between 1825 and 1850 a further 230 names of *Acacia* species were added to the literature.

Expect for Burman's *Adiantum*, the first Australian species were placed in the genus *Mimosa*. Common widespread species *Acacia decurrens*, *A. longifolia*, *A. myrtifolia* and *A. saligna* began in *Mimosa*. Philip Miller had recognised *Acacia* as being different from *Mimosa* in 1754 but until early in the nineteenth century most botanists preferred to interpret *Mimosa* in rather a broad way to include *Acacia*, *Leucaena* and other genera.

There have been about 900 names used for Australian species of *Acacia* so far and the naming continues. Undescribed species are still being found and the naming process will probably go on at least for another decade. There are obviously more names than species but this is not unusual in a large genus. Many botanists of many nationalities have described species since Burman but (in my estimation) three stand out – George Bentham, Ferdinand von Mueller and Joseph Henry Maiden.

Bentham never visited Australia but worked entirely from dried plant specimens and collectors' notes. In a botanical career which continued until he was more than eighty years old, he described many species from Australia and in a series of papers from 1842 to 1875 developed a classification of *Acacia* which is the basis of the one currently accepted. Mueller who laid the foundations for botanical studies in Australia travelled and collected extensively, employed other collectors, sent specimens and notes to Bentham, and described many species himself. Maiden also described many species of *Acacia*, chiefly from northern and Western Australia. Though he was the undisputed expert on the genus for a quarter of a century he might be criticised for merely following Bentham's classification without modification or comment.

Bentham recognised that species of *Acacia* throughout the world fell into six groups. This was not seriously questioned for almost a century but in the last ten years intensive work on pollen, seedling development and chemical constituents of seeds, gum and wood, suggests that there are three major groups, the subgenera mentioned above. Bentham's groups fit fairly well into these broader groups, the main difference being that Australian species such as *Acacia pulchella* and *A. decurrens* with true leaves are more closely related to phyllodinous species than Bentham thought and all form a rather tightly knit 'Australian' group.

Classification of the Australian group is still uncertain. There are two major groups — species with phyllodes with a single nerve (uninerved, see colour plate 7) to which species such as *Acacia terminalis* and *A. decurrens* are related, and species with plurinerved phyllodes with flowers in heads or spikes (see colour plates 21 and 18, together with several well defined smaller groups. *Acacia lycopodiifolia* and *A. alata* (see colour plate 8) are representatives of two of the smaller groups. Unfortunately, no one has yet presented a classification of all Australian species.

I have tried to give an overview of *Acacia*. Much has still to be done. The armchair botanist can speculate as to why *Acacia* should have diversified to such an extent in Australia or why there should be so many species in the south-west. The herbarium botanist can go on describing species, untangling problems of nomenclature and proposing systems of classification. But it will be largely up to the amateur botanist, the naturalist, to broaden our knowledge of *Acacia* by looking at plants in the field and in the garden. Except for widespread economically important species such as Mulga and Brigalow, little is known about the biology of *Acacia*. Growth rhythms, flowering periods, pollination mechanism, seed dispersal, all these things would repay study.

Basic to any study is correct identification. The observer must know the identity of the plant he is observing so that he can communicate his results accurately to others. This book will therefore be of great value to everyone interested in any aspect of the study of *Acacia* for it makes it possible to identify such a large number of species.

Les Pedley
Assistant Director, Queensland Herbarium

The Propagation of Acacias

Growing from seed

With few exceptions, acacias produce hard-coated seeds which need some treatment before sowing to ensure germination. Standard methods are soaking in near boiling water for from thirty minutes to overnight and planting those seeds which have swollen, and re-treating those which have not. Other aids are penetration of the seed-coat by nicking with a sharp blade without damaging the soft inner part of the seed; by rubbing the seed between two pieces of sandpaper or by treating it with acid. The treated seeds are then placed in a free-draining, moist, gravel-loam mix in a plastic pot or seed box, covered to about their own depth with soil and firmed down.

Often it is recommended that soil be sterilised before use to ensure freedom from attack by soil parasites.

The seed boxes or pots, after sowing and watering, should be placed in a warm, sheltered position where early morning sun shines, or in a glasshouse. The seed containers should not be allowed to dry out. Germination time should range from a few days to several weeks if the seed is viable. The seedlings should be potted on while they are still small into plastic pots or bags of suitable size. Use an even mix of river sand or gravel, loam and peat moss or similar material. Repotted plants should be watered and held in a shade house or shady conditions for a few days or even weeks, to recover from the shock of transplanting, before being placed in the open.

Results of experiments suggest that *Acacia* seed may experience a period of dormancy which may vary from one species to another, and that germination may be lower as the seed ages, or that a germination inhibitor may be present.

For those who wish to grow acacias from seed and who do not belong to the society where seed is available from group seed banks, a limited range of seed can be purchased from commercial seed suppliers in all states.

Cuttings

Growing from cuttings is a successful means of propagating good forms of many plants, including acacias, and this method of propagation is being used increasingly by growers. The cutting-grown plant usually will flower much earlier than one grown from seed.

The cutting material may be of firm tip growth or of semi-hardened side-shoots which may be taken at any time that material becomes available, although late summer and early autumn are considered the best times for taking cuttings of many species. Cutting material should be cut cleanly just below a node or leaf joint, or a piece taken with a heel. It is necessary to remove the leaves or phyllodes from the lower two-thirds of the cutting, taking great care not to damage or strip the bark. It is best to cut the leaves off. With larger-leafed cuttings, the leaves should be reduced in size by about two-thirds and any flower buds present should be removed. Before planting, cuttings are often dipped in a hormone powder to stimulate root development.

A suitable medium for striking cuttings is a free draining mix with a good proportion of gravel and some loam in it to retain moisture. Plastic pots or 'gold-lined' cans are suitable containers; terra cotta pots are considered unsuitable because of their tendency to dry out too quickly. It is necessary to make a hole in the mix with a dibber stick before planting each cutting to ensure that no damage is done to the bark. Pots should be watered and placed in a glasshouse; or they may be covered with a plastic bag supported by bent wires or put into a cold frame which may be purchased from any garden supplier. A simple cold frame may be made from a wooden box with sloping sides, which is covered by a plastic-covered hinged lid. The pots or cold frame should be placed in a warm sheltered position and the cuttings should not be allowed to dry out.

It is advisable not to crowd too many cuttings into one container as 'damp-off' may become a problem. It is necessary to check the pots regularly and remove all dead leaves or failed cuttings to maintain cleanliness and avoid the possibility of having to use dangerous fungicidal sprays for control.

The time taken for roots to develop will vary from some weeks to many months depending on the type of cutting material used, the warmth available or the time of year the cuttings were taken. When roots have formed, the new plants should be potted on with care, without damaging the roots, into a soil similar to the cutting mix. The pots should then be watered and placed in a shady position for a few weeks before they are moved to a sheltered, sunny position in the open to grow on to a suitable size for planting in the garden.

Acacias usually grow best in warm, well-drained positions in the garden in full sun or partial shade although a few exceptions thrive in damp situations. Acacias require little fertiliser; an occasional application of blood and bone is beneficial. Light pruning after flowering is recommended to help retain the shape of the bush and perhaps prolong its life.

Buying plants and collecting propagating material

For those who wish to buy their plants, *Acacia* species are available in increasing variety from general and native plant nurseries throughout Australia. Do not be tempted to buy large plants in small containers; their root systems will be too far advanced for successful transplanting.

Those who wish to collect their own propagating material must observe state laws for the protection of plants. These laws can be checked with National Park authorities or Forestry Commissions in all states.

When cutting material is taken, it is imperative that it be placed in a plastic bag, sealed immediately and placed in a cooler, a car refrigerator, or at least in the coolest place possible. *Do not leave in the sun.*

Transplanting directly from the bush, even if local regulations allow it, is not recommended as, apart from depletion of the bush stock, too often the transfer is followed by the death of the plant. An apparently small plant may have a long root system which, when dug up, may suffer severe damage from which the plant may never recover.

Occasionally with a plant which suckers, and quite a few acacias do, it is possible to take a very small sucker and treat it as a cutting. This will grow on successfully.

However, our greatest sense of achievement and pleasure comes from growing our own plants from either seeds or cuttings and sharing the results with friends and other native plant enthusiasts.

Abbreviations

Qld	Queensland
NSW	New South Wales
ACT	Australian Capital Territory
Vic.	Victoria
SA	South Australia
WA	Western Australia
NT	Northern Territory
Tas.	Tasmania
±	more or less
c.	approximately
ssp.	subspecies
var.	variety

Where descriptions of size or number are given in the text, variations from the norm are indicated by a figure in brackets thus: (−6) or (6−). If this is placed *before* the main measurement or number, e.g. (6−) 7·5−14 mm × 0·6−1·5 mm, then it should be understood to mean that the phyllodes (in this case) vary in size between 7·5 and 14 mm in length but have been noted as short as 6 mm. If placed *after* the usual measure it would mean 'as great as'. Length is given before width.

Where the figures refer to flowers, e.g. (15−) 20−35 flowers, the same principle applies.

Key to the Groups

How to use the key

Species with similar characteristics have been grouped together. Holding the flowering specimen in one's hand, if leaves are reduced to phyllodes or are absent, start with the first '1'; if leaves are bipinnate or feather-like, turn to the second '1'.

Assuming that your specimen has phyllodes and has flowers in globular heads on single simple stalks, use the first 'a', then check through the different 'b' and 'c' groups, which outline different types of phyllodes.

If your specimen has flowers in globular heads in racemes, turn to the second 'a' and if the phyllodes are flat with one nerve, the same procedure follows. The 'c' gives further phyllode details and the 'd' bud characteristics.

However, if your specimen has flowers in globular heads on simple stems and phyllodes with several nerves you turn to the next 'a' and so on.

The 'a' gives the type of flower arrangement, the 'b' gives the general phyllode type, the 'c' takes the phyllode description a step further and the 'd' gives a definite flower identification detail.

1 **Leaves reduced to phyllodes or absent**
  a *Flowers in globular heads (rarely slightly elongated)*
    on simple stalks, singly, in clusters, occasionally
    on short or very reduced racemes.

	b	Phyllodes flattened at base into wings continuous with stem	Group 1	1–2
	b	Phyllodes absent, small or with base continuous with stem; branchlets mostly ending in spines	Group 2	3–8
	b	Phyllodes stiff, flat, rounded or ± four-angled, pungent or hard-pointed; one nerve, occasionally more	Group 3	9–19
	b	Phyllodes ± triangular or rounded; one or several nerves	Group 4	20–25
	b	Phyllodes tiny, fine, whorled around stem	Group 5	26–29
	b	Phyllodes with one nerve		
		c Phyllodes fine, linear 2–20 cm long, 1–2 mm wide, occasionally wider	Group 6	30–34
		c Phyllodes flat, small to large, obliquely oblong to broadly lance-shaped 0·6–13·5 cm long, 1–25 (–35) mm wide	Group 7	35–48

a *Flowers in globular heads in racemes*
  b Phyllodes with one nerve, flat.
    c Phyllodes often glaucous, oblong-elliptical,
      oval or occasionally triangular 0·6–10 cm
      long, 2–20 (–45) mm wide   Group 8   49–63
      d Buds enclosed in conspicuous bracts   Group 9   64–65
    c Phyllodes linear, straight to broadly lance-
      shaped 2·5–20 (–27) cm long, (5–) 15–25
      (–85) mm wide; occasionally with more than
      one nerve or with spiny stipules   Group 10   66–85

a *Flowers in globular heads on simple stems*
  b Phyllodes with several nerves
    c Phyllodes stiff, pungent or hard-pointed
      0·6–12·5 (–14) cm long, 1–6 (–15) mm wide   Group 11   86–96
    c Phyllodes narrow, oblong-oval to broadly lance-
      shaped, blunt-tipped, often with conspicuous
      net-like veins 0·4–4·5 cm long,
      1–10 (–15) mm wide   Group 12   97–101

a *Flowers in globular heads in racemes or clusters*
  (occasionally solitary)
  b Phyllodes flat with several nerves, oblong, narrow
    to lance-shaped or elliptical 1·7–15 (–30) cm
    long, 2–35 (–45) mm wide   Group 13   102–110

a *Flowers in spikes*
  b Phyllodes with several nerves
    c Phyllodes flat, stiff, pungent-pointed (base
      continuous with stem in number 112
      A. triptera)   Group 14   111–112
    c Phyllodes fine, linear, round or flat 3–15 cm
      long, 0·7–1 (–2) mm wide   Group 15   113–116
    c Phyllodes flat oblong, oval to rounded, narrow
      to broadly lance-shaped, sometimes blunt-
      pointed 1–25 cm long, 1–95 mm wide   Group 16   117–136

1 **Leaves bipinnate**   Group 17   137–150

1 *Acacia glaucoptera*
Benth.

Common name	Clay Wattle or Flat Wattle
Meaning of name	Referring to blue-green winged stems.
Distribution	WA south-west in Stirling, Warren and Eyre districts from Stirling Ranges to Esperance, in gravel or clay soils on heath or in mallee scrubs.
Habit	Much-branched, sometimes straggling, thick-foliaged shrub 0·3–1·5 m × 1–2 m with flattened, stiff, zig-zagging stems. New growth often bright red or bronze.
Foliage	Flat, thick, blue-green phyllodes, ± suppressed, running into flattened, zig-zagging stem forming overlapping, erect, roughly triangular alternate wings; vein prominent, margins thickened, narrowing into a sharp, often hooked point; stipules small, stiff but not spiny.
Flowers	Dense, bright yellow balls 5–9 mm diameter, each of 30 or more flowers, on usually red, smooth stalks 4–5 mm long, sometimes longer, singly or in pairs, arising from nodes on the midribs. Flowering August–November.
Pods	Stalked, dark brown with wrinkled surface, narrow, curved or twisted, 1–3 cm × 2–3 mm, margins nerve-like, raised over and constricted between seeds.
Seeds	Black, shining, oval 2–3 mm × 1·5–2 mm; seed-stalk fine, short, thickening into a pale cap-like aril.
Identification	Thick, blue-green, flattened stems forming alternate wings.
Comments	Long-lived ornamental shrub which is reported to grow quickly in favourable conditions. It is widely grown in eastern states, but is a little slow in Tas. Some pruning after flowering is recommended to encourage red new growth. Strikes readily from cuttings.

Acacia glaucoptera

2 *Acacia alata*

R. Br.

Common name	Winged Wattle
Meaning of name	Referring to winged branches.
Distribution	WA south-west in Irwin, Darling and Warren districts, usually in moist or shady places, often on river banks, from New Norcia to southern coast; a few areas north of Geraldton (var. *biglandulosa*).
Habit	Many-stemmed, sometimes straggly, green foliaged shrub 0·5–2·5 m × 1–3 m with flattened zig-zagging, winged, usually hairy stems. New growth often reddish or bronze in colour.
Foliage	Phyllodes ± suppressed, appearing as part of the flattened winged stem, erect or spreading with a prominent nerve, tapering into a long spiny point; gland bearing angle(s) on upper margin; spiny stipules at node.
Flowers	Cream or golden yellow balls 7–10 mm diameter, each of 3–12 flowers, on slender stalks 5–10 mm long, singly or in pairs at the nodes. Flowering June–October.
Pods	Brown, usually hairy, curved, oblong 4–7·5 cm × 8–10 mm with thickened margins, rounded over seeds.
Seeds	Dark brown, oval 3–4 mm × 2–3 mm, transverse in pod; seed-stalk folded several times, the last fold thickened into a small aril.
Identification	Flattened, winged stems, spiny stipules, few-flowered flower-heads; var. *biglandulosa* has two, sometimes three gland-bearing angles.
Comments	Semi-hardy shrub which is grown widely in eastern states as an undershrub in cooler districts. Grown for its unusual foliage, it tends to die back under cultivation and requires pruning after flowering. Grown easily from cuttings.

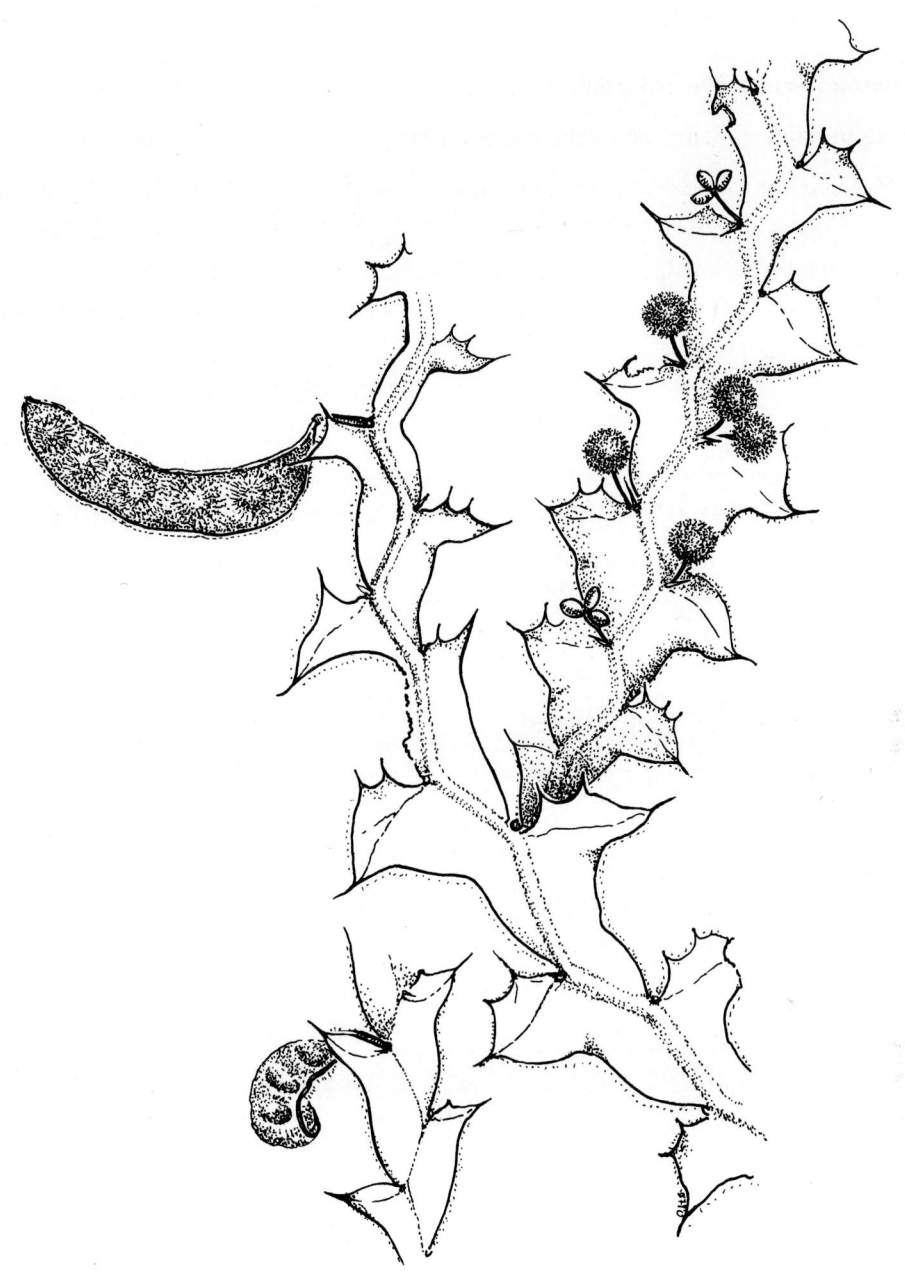

Acacia alata

3 *Acacia spinescens*

Benth.

Common name	Spiny Wattle
Meaning of name	Referring to branchlets ending in a spine or sharp point.
Distribution	Usually found in sandy dry areas in many parts of SA, north-west Vic. and western NSW.
Habit	Small, usually leafless, rigid shrub mostly less than 1 m tall with sturdy, round, smooth, finely ribbed stems, mostly ending in spines.
Foliage	When present, phyllodes are narrow, tapering to base 1–5 cm × 2 mm with a curved point; small brown deciduous scales often remain where phyllodes have fallen.
Flowers	Bright yellow balls 5–6 mm diameter, each of 3–8 flowers, stalkless or on stalks 3–6 mm long along the stem. Flowering July–October.
Pods	Small, dark brown, bead-like, curled and twisted, 2–3 cm × 2–3 mm with slightly thickened margins, rounded over and constricted between seeds.
Seeds	Tiny, shiny, black, oval *c.* 2 mm × 1·5–2 mm, longitudinal in pod; short seed-stalk thickened into a club-like aril.
Identification	Spiny branchlets, usually no phyllodes and few-flowered, almost stalkless or stalkless ball flowers.
Comments	Will tolerate a little shade but requires a warm, well-drained position; it grows successfully in most southern states including Tas., but it is cut occasionally by late heavy frosts. Grown from cuttings and seeds.

Acacia spinescens

4 *Acacia continua*

Benth.

Common name   Thorn Wattle

Meaning of name   Referring to continuation of phyllodes down stem.

Distribution   Found in rocky, dry regions of inland SA, NT and in Broken Hill area, NSW.

Habit   Small, erect, stiff, sometimes rounded shrub usually under 1 m tall, sometimes a little taller, with green, round, striated stems becoming brown when older.

Foliage   Bright green, round, often recurved phyllodes 0·8–3 cm long, narrow with some longitudinal nerves, continuous with stem at base and tapering at tip into a long sharp point; phyllodes longest on lower part of shrub; gland, if present, about halfway along top margin.

Flowers   Large, deep yellow balls 8–10 mm diameter (sometimes slightly elongated), each of 15–30 flowers, on very short smooth stalks, solitary or in pairs; buds encased in conspicuous brown bracts. Flowering August–October, sometimes earlier.

Pods   Dark brown, much-curved or twisted, 4–8 cm × 3–4 mm, raised over and evenly constricted between seeds, ending in a long thin point.

Seeds   Dark brown with paler centre, oval 3–3·5 mm × 2–2·5 mm, longitudinal in pod; short almost straight seed-stalk ending in a thickened aril.

Identification   Phyllodes continuous with stem, large and almost stalkless ball flowers, and buds in conspicuous bracts.

Comments   Grown from cuttings and seeds; widely grown in most southern areas in full sun or partial shade in a well-drained position. It has been burned by late frosts in Melbourne, but is growing successfully in Tas.

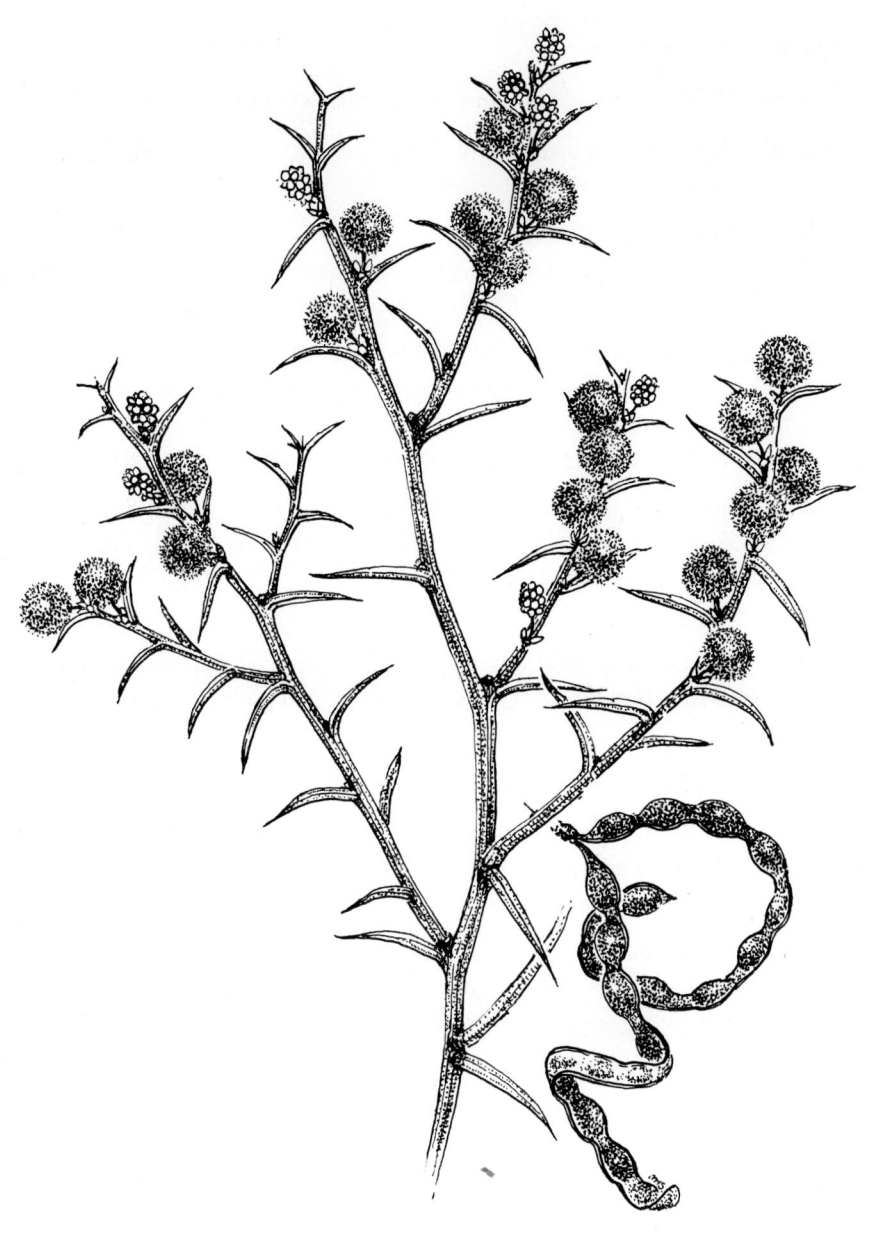

Acacia continua

5 *Acacia daviesioides*

Gardn.

Common name   None known.

Meaning of name   With foliage like a *Daviesia*, a member of the pea family.

Distribution   WA in Irwin district, e.g. in Latham, Buntine and Wubin areas, on sand or gravel.

Habit   Rigid, much-branched shrub 0·3–1 m × 0·6–2·5 m; branches ascending, twisted, round and prominently ribbed, somewhat resinous; branchlets spinescent, occasionally zig-zagging at tips.

Foliage   Apparently 'leafless', but phyllodes are thorn-like, much recurved, ± four-angled in cross section, 5–7 mm long, widest at base and running into stem as a rib; tapering at tip into a sharp curved point; a raised circular gland occasionally on upper surface between two nerves.

Flowers   Large, bright yellow balls 8–10 mm diameter, each of 8–15 (–20) loose flowers, on hairless stalks 5–9 mm long, usually singly in the leaf axils. Flowering July–September.

Pods   Dark reddish-brown, firm textured, slightly curved, 5–7 cm × 3·5–4 mm, with nerve-like margins, alternately rounded over but little constricted between seeds. Immature pods are wine red.

Seeds   Dark brown, oval, 2·5 mm × 1·5–2 mm, longitudinal in pod; seed-stalk short, thickening into a pale club-shaped aril.

Identification   Thorn-like phyllodes, prominently ribbed stems ending in spines, large flower-heads of few flowers. Some relation to *A. volubilis* but it differs in sharply acute phyllodes and more numerous ribs of stems.

Comments   An unusual feature shrub for a well-drained, warm position in a rockery.

Acacia daviesioides

6 *Acacia wiseana*

Gardn.

Common name	None known.
Meaning of name	Named after F. J. S. Wise, premier of WA (1945–1947).
Distribution	WA in Austin, Ashburton, deserts of Carnegie district, on red, sandy loam, on river banks, clay pans and edges of samphire flats, common in dense scrub, often with *A. sclerosperma*; occurs in NT.
Habit	Intricately branched, almost leafless shrub 1–3 (–5) m × 3–5 m with slightly flaking, grey-brown bark on main stems; branchlets green, prominently and finely ribbed, spinescent; spines lateral, finely lined 2–4 cm long.
Foliage	Few, narrow, short-lived phyllodes 1 cm × 1 mm, central nerve ending in straight, sharp, not pungent point; gland towards middle of top margin; bracts long-pointed.
Flowers	Bright yellow balls 8–10 mm diameter, each of about 20 flowers, on slender, hairless stalks 6–10 mm long, solitary or in small ±6-flowered racemes. Flowering August–September.
Pods	Large, yellowish-brown, firm textured 7–15 cm × 8–12 mm with thickened, paler margins, much rounded over and constricted between seeds. Immature pods red.
Seeds	Dull, brown-black, round 6–9 mm diameter, seed-stalk thickening into a yellowish, turban-like aril.
Identification	Finely ribbed, spinescent branchlets and lateral spines; very large pods and large round seeds.
Comments	No information available on cultivation.

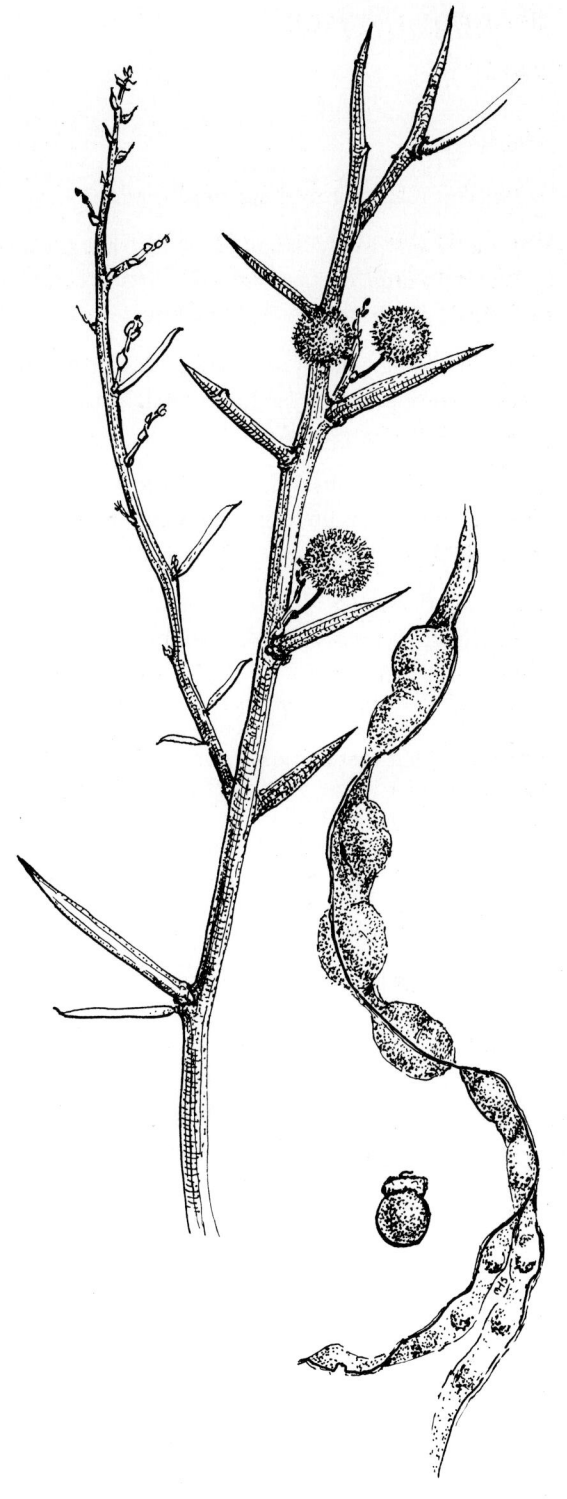

Acacia wiseana

7 *Acacia ulicina*
Meisn.

Common name	None known.
Meaning of name	Gorse-like; referring to shape of phyllodes.
Distribution	WA south-west in Irwin and Avon districts from near Geraldton to Kellerberrin area, on sandy or gravelly soil often in eucalypt woodland, sometimes growing with *A. lasiocarpa*.
Habit	Rigid, spreading, prickly shrub 0·2–2 m × *c.* 1 m with finely ribbed grey-brown branches; branchlets many, small, spreading, ending in spines or reduced to thorns.
Foliage	Grey-green, linear phyllodes 7–15 mm × 1–1·5 mm crowded at tips of spiny branchlets, up to 25 mm long on main branches; mid-nerve ending in a long, oblique point.
Flowers	Small, yellow balls 5–7 mm diameter, each of about 20 flowers, on usually fine, hairless stalks 4–10 mm long, singly in axils. Flowering August–September.
Pods	In clusters, dark brown, bead-like, curled or twisted, 3·5–5 cm × 2–3·5 mm, alternately raised over and much constricted between seeds.
Seeds	Black, shining, oval, almost longitudinal in pod; seed-stalk slender, thickened into a pale club-shaped aril, almost as long as seed.
Identification	Crowded phyllodes, spiny branchlets, bead-like pods.
Comments	It is being grown successfully in eastern states; requires a position in full sun.

28

Acacia ulicina

8 *Acacia acanthoclada*

F. Muell.

Group 2

Common name	Harrow Wattle
Meaning of name	Referring to thorny or prickly young shoots or branches.
Distribution	Scattered throughout the dry, sandy, inland plains or low open woodlands of SA, south-west WA and, occasionally, in the north-west mallee area of Vic. and south-west NSW.
Habit	A small, rigid, much-branched, spiny shrub, 0·3–2 m tall, up to 1 m wide with grey bark and round branchlets covered in dense or scattered white hairs; leafy stems ending in long, fine, hard spines.
Foliage	Phyllodes tiny, usually crowded, oblique, green, hairy, 3–8 mm × 1–4 mm, broadest at tip, ending in a small recurved point; main nerve running close to and parallel with straighter lower margin; narrowing at base into a short stalk; persistent small brown stipules at base.
Flowers	Very bright yellow balls, 6–7 mm diameter, each of 25–30 flowers, on smooth solitary stalks, 4–10 mm long. Flowering August–October.
Pods	Smooth, dark brown, narrow, flat pods, coiled and twisted, about 3 mm wide, a little constricted between seeds.
Seeds	Brown, oval-oblong, longitudinal in pod; seed-stalk short and thickened into a boat-shaped aril.
Identification	Small, rudder-shaped phyllodes, spiral pods and branchlets ending in spines.
Comments	Grown from seeds; it requires a warm, well-drained position in full sun or a little shade. It grows best in sand or loam but will tolerate some clay, and it is considered to be quite long-lived. It is growing satisfactorily in Melbourne. It has been listed as 'rare and needs constant monitoring' in SA and as an endangered species in Vic.

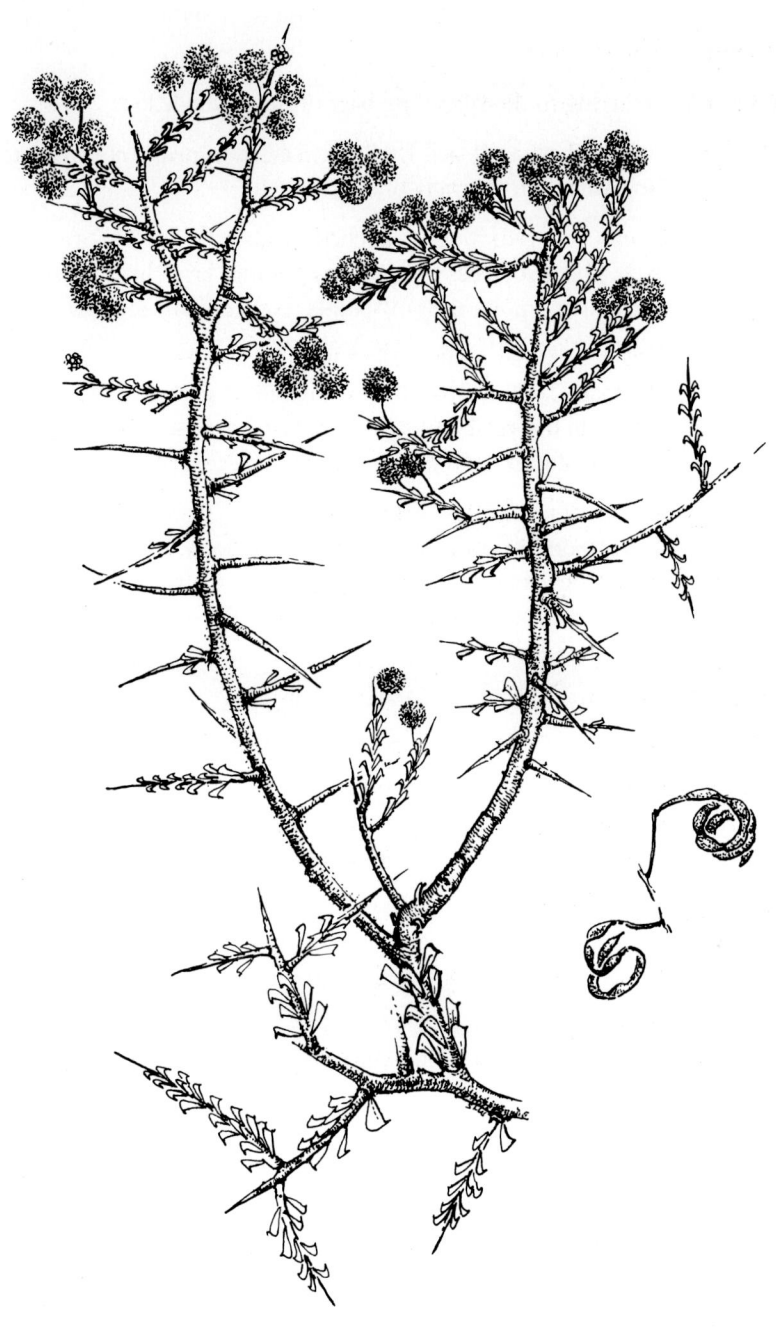

Acacia acanthoclada

9 *Acacia pachypoda*

Maslin

Common name	None known.
Meaning of name	Referring to the spreading base of the phyllode.
Distribution	WA in Coolgardie and Eyre districts, from near Coolgardie to south of Norseman, on a variety of soils.
Habit	Prickly foliaged, much-branched shrub 0·6–1 m × 0·6–2 m with grey bark; upper surfaces of branches peeling; branchlets smooth, round, sometimes ending in spiny points. New growth often wine red.
Foliage	Green, rigid, straight or slightly curved, usually spreading, round phyllodes 7–26 mm × 1 mm, smooth or slightly wrinkled, nerveless, ending in a very sharp point; spreading at base, leaf stalk absent; small gland 1–2 mm from base on upper surface; stipules deciduous.
Flowers	Masses of pale yellow balls 3–5 (–8) mm diameter, each of about 8 flowers, on fine hairless stalks 4–7 mm long, in much reduced racemes of 2–3 flowers, often growing out into a leafy shoot. Buds emerging from conspicuous, smooth, brown, deciduous bracts. Flowering August–September.
Pods	Grey-brown, linear, curled when dry, 4–6 cm × 2–3 mm, raised over and slightly constricted between seeds. Young pods dull wine colour.
Seeds	Brownish-black, shining, oblong, 3·5–4 mm × 1·5–2 mm, longitudinal in pod; seed-stalk thread-like, tiny, abruptly thickened into a cap-like aril. Seeds mature in mid-December.
Identification	Slightly resembles *A. colletioides* and *A. nyssophylla*, but differs in its nerveless phyllodes with spreading base and lack of leaf stalk; much reduced raceme of pale yellow flowers; large brown bracts encasing buds.
Comments	No information available regarding cultivation.

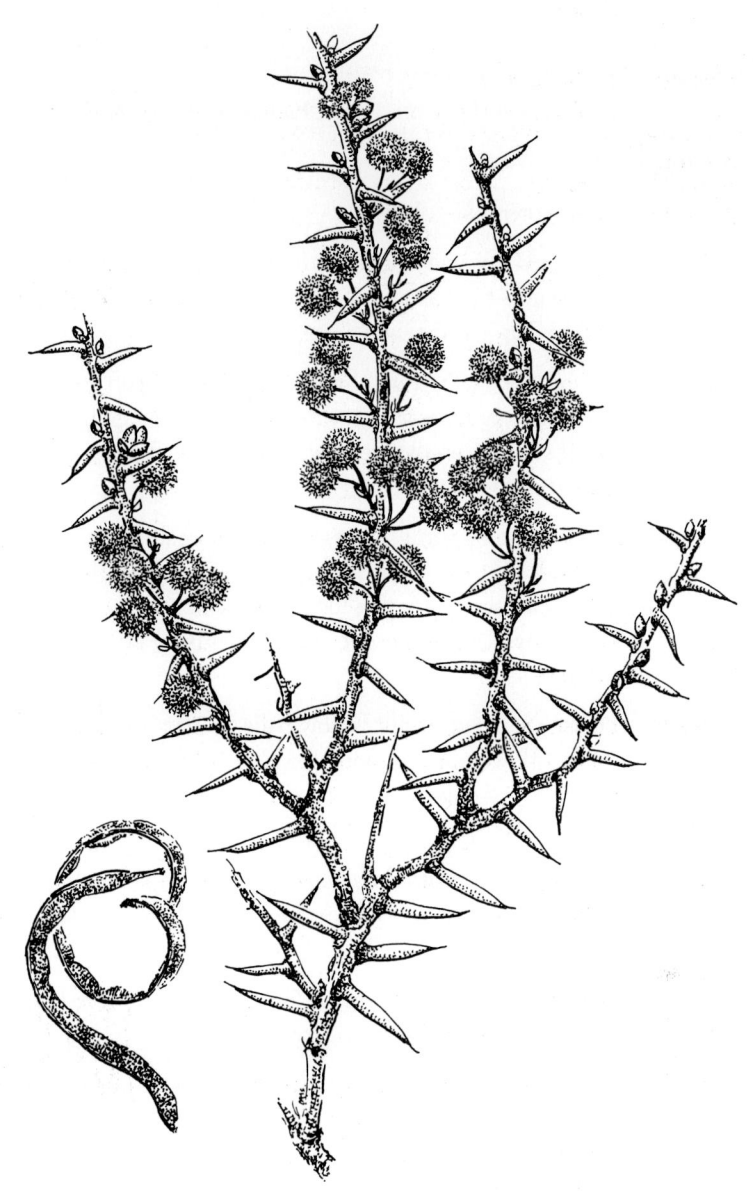

Acacia pachypoda

10 *Acacia ulicifolia*

(Salisb.) Court

Group 3

Synonyms	*A. juniperina* (Vent.) Willd. *A. brownei* (brownii) (Poir) Pedley in *Austrobaileya* 1, 3, 1979.
Common name	Prickly Moses
Meaning of name	With gorse-like foliage.
Distribution	Scattered but widespread through dry eucalypt woodlands, some ranges and heaths of coastal and near coastal regions in all eastern states from Qld to Tas.
Habit	A variable, rigid, much-branched prickly shrub, prostrate to 2 m tall with hairy or smooth, round, sometimes drooping branches; persistent small stipules on old stems giving twiggy, rough appearance. Young growth hairy.
Foliage	Prickly green, scattered or crowded, spreading rigid somewhat flattened phyllodes (6–) 7·5–14 mm × 0·6–1·5 mm with prominent nerve on each side tapering into a long, very sharp point; swelling slightly at base; tiny gland sometimes present causing an indentation in upper margin.
Flowers	Dense cream or bright yellow balls 8–10 mm diameter, each of (15–) 20–35 flowers, on smooth, slender, solitary stalks (5–) 8–15 mm long; bracteoles sometimes obvious in unopened buds. Flowering July–September.
Pods	Brown, linear, flattened, usually curved 2·5–6 cm × 3–5 mm with lighter nerve-like margins, raised over, evenly constricted between seeds.
Seeds	Dull black, oval to oblong 4–5 mm × 2·5–3 mm, longitudinal in pod; threadlike seed-stalk a little folded.
Identification	Closely allied to *A. echinula* DC but differs in phyllode shape, gland and smaller pods. *A. ulicifolia* var. *brownei* is recognised by some authorities as a separate species, *A. brownei*. It is a decumbent shrub to 0·4 m high with phyllodes to 1·5 cm long and bright yellow flowers.
Comments	Grown from cuttings and seeds. It is a hardy shrub which is grown successfully in gardens from Rockhampton in Qld to Tas.

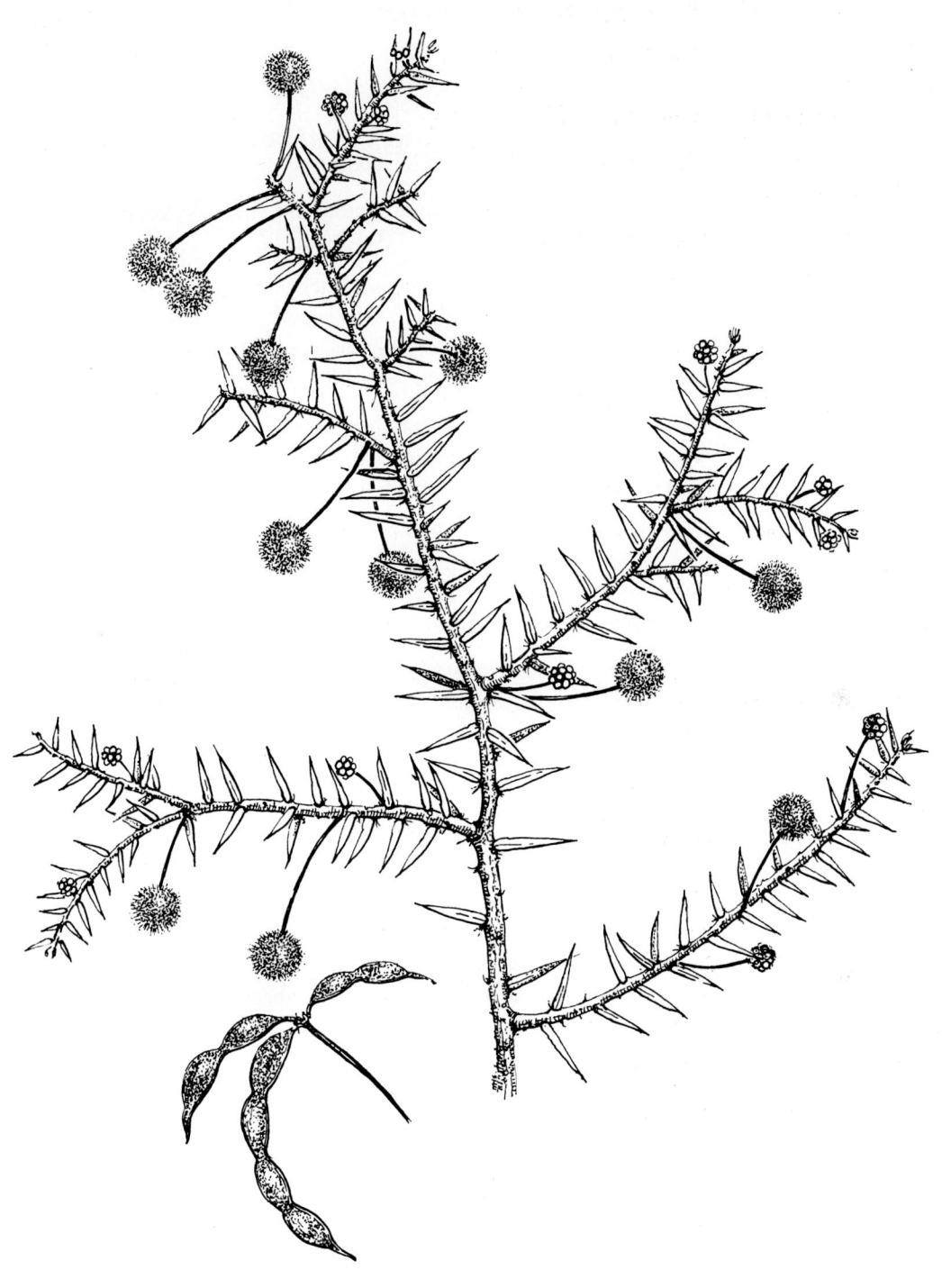

Acacia ulicifolia

11 *Acacia aculeatissima*

Macbride

Group 3

Common name	Thin-leaf Acacia
Meaning of name	Referring to prickly, pointed phyllodes.
Distribution	Widespread, sometimes common; often in hilly, rocky country in Vic. and southern NSW.
Habit	Variable, much branched, spreading, pickly shrub, usually prostrate to 0·5 m tall, occasionally taller, 1–2 m across with reddish-brown smooth or hairy branchlets. Some varieties have densely crowded phyllodes.
Foliage	Prickly, bright green, stiff, straight phyllodes 0·5–2 cm × *c.* 1 mm, usually sloping back down stem, but not always; a raised nerve on each side tapering into a fine sharp point. Small stiff stipules at base.
Flowers	Pale or bright yellow balls 7–9 mm diameter, each of about 20 flowers, on slender, usually smooth stalks up to 15 mm long, singly, or occasionally two or three together in leaf axils. Flowering August–October, sometimes later.
Pods	Reddish-brown, smooth, straight or curved 2–6 cm × 3–4 mm with slightly thickened margins, rounded over, little (if at all) constricted between seeds.
Seeds	Oval, dull black, 3·5–4 mm × 2 mm, longitudinal in pod; yellowish seed-stalk folded several times and thickened into a small cap-like aril.
Identification	Sharp pointed phyllodes usually pointing down stem; often pale flowers on long slender stems; thickened yellowish seed-stalk.
Comments	Hardy shrub which may grow 1·6 m or more tall under cultivation; suited to a variety of soils as long as well-drained; prefers semi-shade. Prostrate form is an ideal rockery or ground cover plant. Grown from seeds or cuttings.

Acacia aculeatissima

Benth.

Common name	None known.
Meaning of name	Without stalks.
Distribution	WA south-west in Irwin, Avon, Stirling and Coolgardie districts, e.g. Shark Bay area, near Geraldton and south-west of Perth; common in sand and open heathlands.
Habit	Rigid, much branched, prickly shrub 1–1·5 m × 0·7–2 m with smooth, sometimes mottled bark; branchlets ± round, often ending as long, thin spines. New growth hairy, crowded at tips, young stems normally covered with white, woolly hairs.
Foliage	Green, scattered, usually spreading, rigid phyllodes 10–18 mm × 0·5–1 mm, sometimes pointing down stem, one, rarely two, prominent nerves tapering into a stiff, very sharp point; brown deciduous stipules at base of young phyllodes.
Flowers	Numerous bright yellow balls 6–8 mm diameter, each of 20–30 flowers, solitary on very short stalks 1–4 mm long, sometimes in long sprays; sepals distinct, linear-spathulate with dark tips. Flowering August–November.
Pods	Dull brown or grey with hairs 1·5–3·5 cm × 3–5 mm raised over seeds, margins thickened, straight, not constricted between seeds. Immature pods white, woolly-hairy.
Seeds	Brown, tiny, oval, longitudinal or slightly oblique in pod; seed-stalk short, thickened into a pale yellow, boat-shaped aril extending half the length of seed.
Identification	Sharp pointed phyllodes, crowded at tips, spines at end of branches, very short-stalked flowers; woolly new growth.
Comments	It is growing successfully in Melbourne in a warm, well-drained position where it grew quickly to 1 m × 1·5 m. Grown from seeds; it is not considered easy from cuttings.

Previously known as *A. sphacelata*.

Acacia sessilis

13 *Acacia rupicola* Group 3
F. Muell. ex Benth.

Common name	Rock Wattle
Meaning of name	Growing in stony or rocky places.
Distribution	Widespread in SA often in sandy or dry stony country, and on Grampians and Mt Arapiles areas of western Vic.
Habit	Prickly, erect, shining, viscid shrub 1–2·5 m × 1–2 m; branches arching, light grey-green, slightly angular or marked with resinous ridges.
Foliage	Rigid, shining green, often sticky, prickly phyllodes 7–25 mm × 1–2·5 mm with prominent central nerve ending in a very sharp long point; swollen and rounded at base before joining stem, no stalk; tiny gland near base.
Flowers	Pale yellow balls 6–8 mm diameter, each of 20–25 flowers, on solitary, smooth, sometimes sticky stalks 5–15 mm long. Flowering August–January, sometimes later in the Grampians, Vic.
Pods	Red-brown with net-like veins, narrow, rounded 2·5–8 cm × 3–5 mm straight or curved, slightly if at all constricted between seeds.
Seeds	Brown, oblong, 4 mm × 2 mm, longitudinal in pod; seed-stalk folded several times before thickening into a broad aril.
Identification	Somewhat resembles *A. genistifolia* but differs in shape of phyllodes, arrangement of flowers, pods and seed-stalks.
Comments	Grown from cuttings and seeds. A hardy, frost-resistant shrub useful in rockeries; grown widely in southern states in full sun or partial shade. It has been listed as an 'endangered' species in Vic.

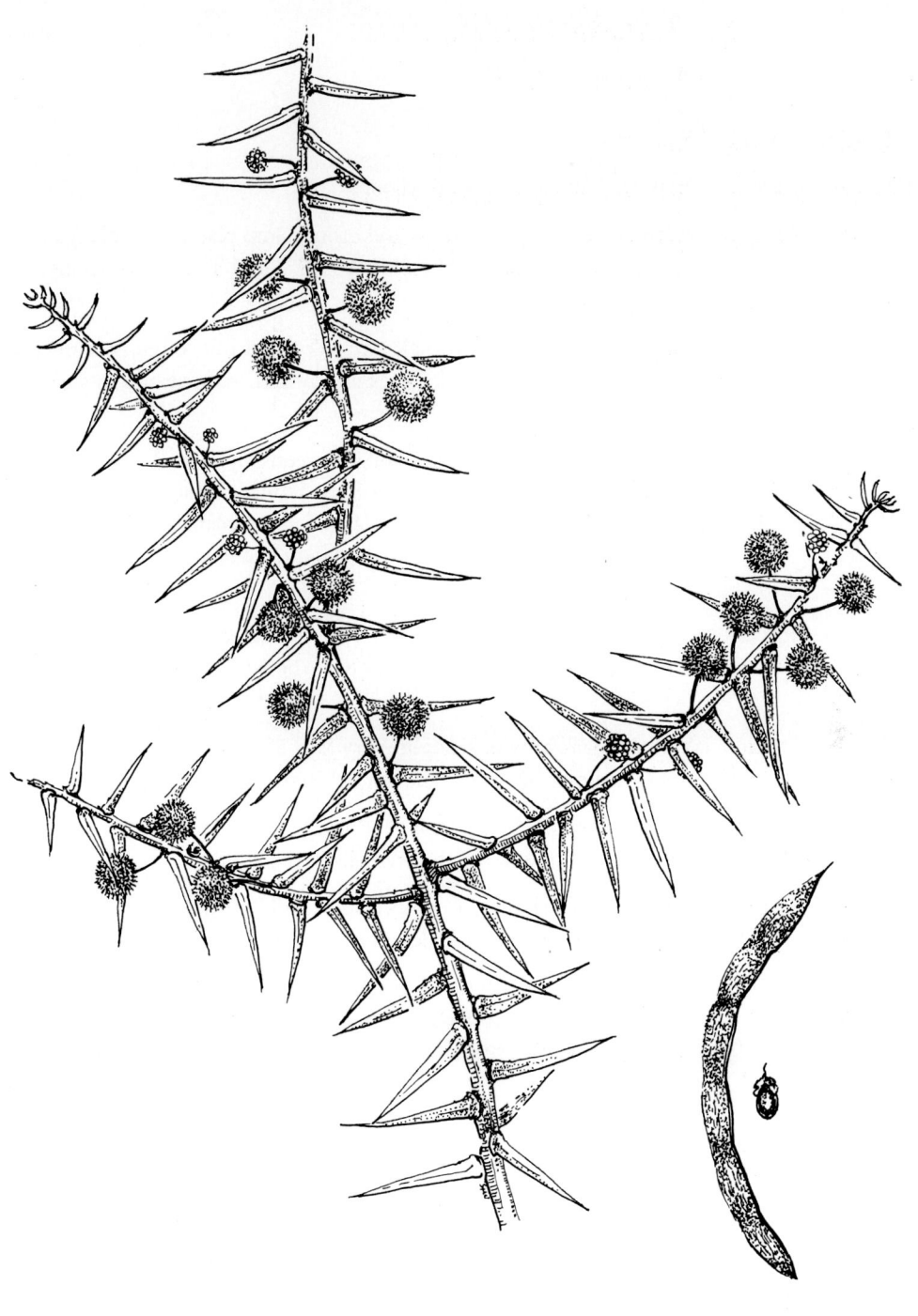

Acacia rupicola

14 *Acacia siculiformis*
A. Cunn. ex Benth.

Common name	Dagger Wattle
Meaning of name	Referring to dagger-like phyllodes.
Distribution	Along creek banks, swamps and other damp places, especially in elevated rocky areas of Vic., NSW, ACT and Tas. at elevations between 1000–1500 m.
Habit	Erect, prickly shrub to small tree 0·5–3 m tall, often procumbent and straggling, with flaky bark and round, reddish, often scaly branchlets.
Foliage	Stiff, linear, usually slightly curved green phyllodes 1–3 cm × 1·5–4 mm with margins sometimes slightly thickened, a prominent central nerve tapering into a long stiff sharp point; narrowing at base into a widened tiny stalk; a small gland usually at base; minute stipules present.
Flowers	Numerous pale yellow balls 5–6 mm diameter, each of about 30 flowers, on little or no stalk, singly at the base of phyllodes. Flowering September–November.
Pods	Stalked, dark brown, straight, very flat, 2–4 cm × 5–7 mm with thickened margins, swollen over and narrowed between seeds.
Seeds	Shiny, brownish-black, oval 3–4 mm × 2–2·5 mm, transverse or oblique in pod with a fine thread-like seed-stalk.
Identification	Somewhat resembles *A. genistifolia* Lindl. but differs in stalkless flowers, more curved phyllodes and transverse or oblique seeds.
Comments	Very hardy; it will grow successfully in damp or rocky cool situations in cool temperate climates and will tolerate frost and snow. Grown from cuttings and seeds.

Acacia siculiformis

15 *Acacia genistifolia*
Lindl.

Synonym	*A. diffusa* Lindl.
Common name	Spreading Wattle
Meaning of name	With Genista (broom)-like leaves.
Distribution	Common in dry sclerophyll forests and heathlands throughout south-eastern NSW, ACT and Vic., with the exception of mallee areas; widespread and abundant in Tas.
Habit	Variable, much branched, often straggly, prickly shrub 1–3 (–5) m tall with smooth grey bark and rigid spreading branches; branchlets angular at first, smooth or scaly, scarred where phyllodes have fallen.
Foliage	Variable, spreading, green, thick, rigid, usually straight phyllodes 1·5–3·5 (–5·5) cm × 1–3 (–6) mm with prominent central nerve and nerve-like margins tapering into a long, very sharp point; slightly widening at base; a gland near base. Phyllodes sometimes vary greatly in size on the same bush.
Flowers	Pale to bright lemon-yellow balls 6–8 mm diameter, each of about 20 flowers, on smooth, usually long stalks 5–20 mm long, usually 2 or 3 together but occasionally up to 6. Flowering July–October, sometimes earlier.
Pods	Stalked, light reddish-brown, a little curved, 4–10 cm × 4–7 mm, with slightly thickened margins, rounded over and little if at all constricted between seeds.
Seeds	Black, oblong 4–5 mm × 2–2·5 mm, longitudinal in pod with white seed-stalk folded several times at end of seed.
Identification	Spiny, usually straight phyllodes with prominent mid-vein; usually 2–3 or more flowers at base of phyllodes.
Comments	Hardy for most situations and grown widely in eastern NSW, Vic. and Tas.; likes a sunny or partially shaded, well-drained position. Will tolerate pruning after flowering. It is grown from cuttings or seeds.
	A brighter flowered form with wider phyllodes to 6 mm wide is found in eastern Tas.; these wider phyllodes often have a second longitudinal nerve near the top margin.

Acacia genistifolia

16 *Acacia axillaris*
Benth.

Group 3

Common name	None known.
Meaning of name	Referring to flowers emerging almost stalkless from leaf axils.
Distribution	Restricted to parts of eastern and central Tas. on slopes and in valleys of St Pauls, Elizabeth and Clyde rivers.
Habit	A straggly but usually dense prickly shrub 2–4 m tall branching from near ground level with smooth grey bark, rough where twigs have fallen; branchlets reddish, with scattered hairs, slightly angular at first. New growth slightly reddish, phyllodes and stems clothed with appressed white hairs.
Foliage	Smooth, green, spreading, straight, sharp-pointed phyllodes 1·5–3 (–5) cm × 1·5 mm with central nerve raised on each side ending in a long fine sharp point; narrowing at base into a small widened stalk; no gland seen.
Flowers	Honey-perfumed pale yellow balls 5–6 mm diameter, each of 2–4 flowers, stalkless or on very short stalks 2–4 mm long, in pairs or small clusters in the leaf axils. Flowering September–October.
Pods	Thin-textured, rusty brown like small dried leaves, 3–4 cm × 2–3 mm with slightly flattened margins, rounded over and a little constricted between seeds, ending in a long point.
Seeds	Shiny black, 3–3·5 mm × 1·5 mm, longitudinal in pod with a fine seed-stalk folded several times at top of seed into a pale cap-like aril.
Identification	Likened to another Tas. endemic *A. riceana* Henslow but differing in shape of buds, stalkless or nearly stalkless, few-flowered ball flowers and phyllodes with a central vein.
Comments	Hardy, suitable for cooler climate gardens with adequate moisture. This *Acacia*, which was not collected for about eighty years, must have been overlooked when not in flower, as it is quite common in its chosen localities. It is not protected in a reserve and is considered vulnerable as it is exposed to clearing, grazing and to fire. It is easily grown from seeds.

Acacia axillaris

17 *Acacia tetragonophylla*

F. Muell.

Common name   Dead Finish, Kurara

Meaning of name   Having four-sided phyllodes.

Distribution   Widespread, sometimes common, in a variety of soils in arid areas of the inland, often where there is some underground moisture in mulga scrubs and plains of Qld, NSW, SA, NT and WA in annual rainfall areas of 150–250 mm.

Habit   Prickly, much-branched, spreading, often straggly shrub or small tree 1·5–4 (–6) m × 1–4 m with fissured grey bark at base of trunk and crooked, round, twiggy, sometimes scaly branches. Phyllodes and buds emerge from the same curly bracts. A form from near Adavale, Qld, is a tree to 6 m tall with pendulous branches weeping to the ground, and phyllodes less sharp and less brittle than usual.

Foliage   Spiny, bright green, slender, distinctly angular, spreading, rigid phyllodes 1–3 (–5) cm × *c*. 1 mm usually clustered 3 or 4 or more together embedded in curly grey bracts; one or two longitudinal nerves on each face ending in a long, very sharp, rigid point.

Flowers   Dense deep yellow balls 1 cm diameter, each of 50 or more flowers, on fine, smooth axillary stalks 10–15 (–22) mm long, singly, in pairs or clusters of up to 5 flower-heads. Flowering influenced by adequate rainfall but usually May–September.

Pods   Conspicuous bunches of dark brown, usually woody, beadlike pods, curled or twisted 3–10 cm × 4–8 mm with thickened yellowish margins, raised over and ± evenly much constricted between seeds. Pods usually remain on bushes after shedding of seeds.

Seeds   Black, ± round, flattened, 4–5 mm × 3 mm, longitudinal in pod; yellow seed-stalk much thickened at base, completely encircling seed. Seeds mature January.

Identification   Clusters of four-angled, very prickly phyllodes, large dense flowers and distinctively curled pods.

Comments   Hardy, drought and frost resistant, reasonably salt-tolerant shrub for warmer climates; recommended plant for Alice Springs area, but not successful in Tas.

   Timber is close-grained, heavy and tough, reddish-brown with pinkish stripes, and smells of violets when first cut.

Acacia tetragonophylla

18 *Acacia gonophylla*

Benth.

Common name	None known.
Meaning of name	Referring to the many-angled phyllodes.
Distribution	WA south-west in Eyre and Stirling districts, e.g. Albany to Esperance areas and further east, on open sand plains in low woodlands.
Habit	Low growing, rigid, much-branched shrub 0·6–1 m × 1–1·5 m with rough grey bark and angular branches.
Foliage	Dull green, stiff, ascending, slightly incurved, angled phyllodes 1–4·5 cm × 1·5–3 mm, often slightly wider towards tip; very prominent longitudinal nerves ending in a short, stiff but not sharp point; tapering at base. Small gland some distance from base.
Flowers	Whitish to pale cream balls 6–9 mm diameter, each of 12–20 flowers, on hairless, sometimes red stalks 6–8 mm long, singly or in pairs in upper axils. Flowering August–November.
Pods	Red-brown, thin textured, curved, linear, flat 4–7 cm × 2–3 (–4) mm, slightly raised over and constricted between seeds. Immature pods often red.
Seeds	Black, oblong, 2·5–3 mm × 1·5 mm, longitudinal in pod; seed-stalk long, last fold ending in a broad cap-like aril.
Identification	Appears to be allied to *A. sulcata*, but differs in the prominently 5-nerved, thick phyllodes and cream flowers.
Comments	Reported to be long-lived in favourable conditions. It is growing successfully in south-eastern states in coarse sand, sandy loam and clay soils. Grown from cuttings and seeds.

Acacia gonophylla

19 *Acacia nigripilosa*
Maiden

Common name	None known.
Meaning of name	Referring to black hairs on the petals.
Distribution	WA south-west in Irwin, Avon and Coolgardie districts, e.g. Wyalcatchem, Morawa or Kellerberrin, often in *Casuarina campestris* and mallee-type scrub, on gravelly or sandy soil.
Habit	Slender, erect shrub 0·5–2 (–3) m × 1–3 m with smooth, grey bark; branchlets smooth, slightly angular; stems sometimes slightly zig-zagging between well-spaced phyllodes.
Foliage	Variable, green, stiff, normally narrow phyllodes (2·5–) 3·5–7 cm × 1–2 (–7) mm with prominent raised mid-nerve, ending in a brown, sharp, hooked or straight point; gland indented at or near base. Broad phyllode forms from north of Latham and near Dowerin are lance-shaped with a sharp, hooked point, 2·5–6 cm × 4–7 mm.
Flowers	Perfumed, bright yellow balls 8–10 mm diameter, each of about 20 flowers, on hairless or occasionally hairy stalks 5–8 mm long, in reduced racemes of 1 or 2 flowers, sometimes growing out into leafy shoots. Buds, somewhat elongated, encased in large, brown, deciduous bracts; petals brown, black hairs on upper half. Flowering August–September.
Pods	Dark brown with rough surface, curved or straight 5–8 cm × 4–6 mm with thickened margins, raised over and regularly constricted between seeds.
Seeds	Black, shining, oval 4 mm × 2 mm, longitudinal in pod; seed-stalk folded and thickened into an irregular fleshy aril.
Identification	Conspicuous bracts when in bud, position of gland, large flowers in reduced racemes.
Comments	An attractive, large-flowered shrub which requires a well-drained position in full sun in the south. It is growing successfully in Vic. in sandy or clay soils.

Acacia nigripilosa

20 *Acacia gunnii*
Benth.

Synonym	*A. vomeriformis* A. Cunn. ex Benth.
Common name	Ploughshare Wattle
Meaning of name	Named for R. C. Gunn (1808–1881), early botanist and collector in Tas.
Distribution	Widespread, occasionally abundant in sandy, gravelly soils mainly on slopes and tablelands in Vic., NSW, ACT, SA, Tas.; very rarely in Qld (near Stanthorpe); often near coast in Tas.
Habit	Small, straggly, often spreading, prostrate or erect prickly shrub to 1 m tall, with slender, round, hairy or smooth branchlets. New growth often densely hairy.
Foliage	Variable, rigid, broadly or narrowly triangular phyllodes 4–10 (–13) mm × 2–5 mm with prominent nerve near lower straight margin and tapering into a long fine sharp point; narrowing abruptly at base; upper margin sometimes rounded or raised into an acute angle often with a gland at the bend; small, stiff, often hairy, persistent stipules.
Flowers	Pale yellow to cream balls 6–7 mm diameter, each of 20–30 flowers, on slender hairy stalks up to 10 mm long, singly, in pairs or clusters; buds are a little like burrs with projecting bracteoles. Flowering July–September.
Pods	Dark brown, smooth or hairy, flat 2–4 cm × 3–6 mm with thickened margins, much constricted and lengthened between seeds.
Seeds	Dull black, oval-round 4 mm × 2·5 mm, longitudinal in pod; short threadlike seed-stalk.
Identification	Small, usually triangular phyllodes, position of gland, single vein near lower margin, pale flowers.
Comments	Easily grown from cuttings or seeds. It is tolerant of a wide range of conditions and grows well in full sun or partial shade in a well-drained site, but needs some moisture. It is ideal for a rockery.

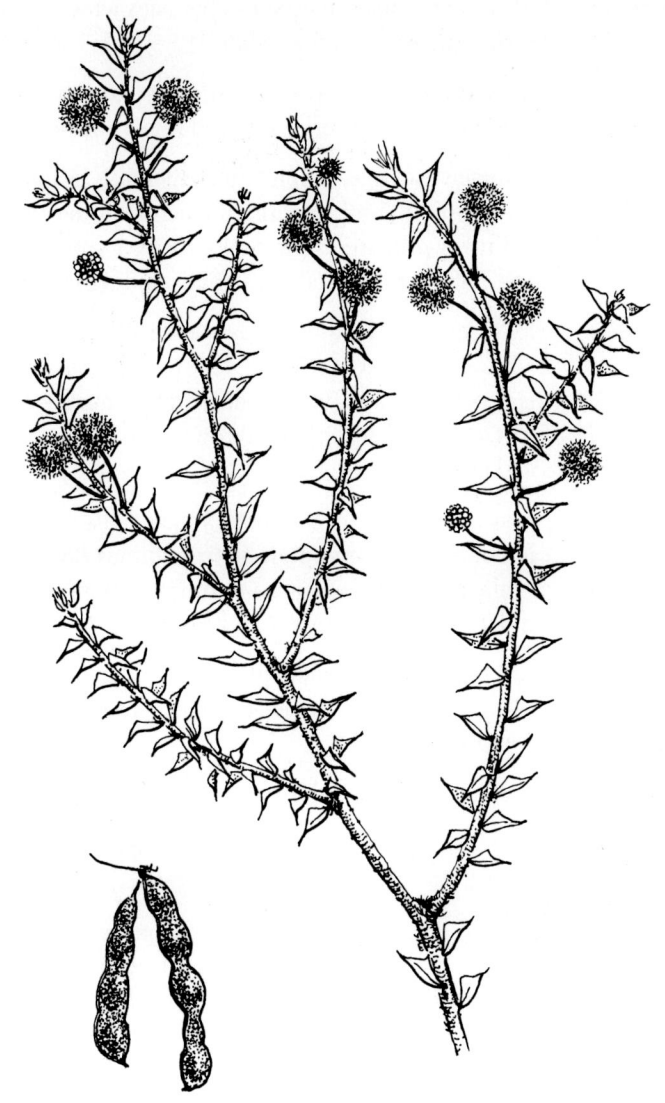

Acacia gunnii

21 *Acacia bidentata*

Benth.

Group 4

Common name	None known.
Meaning of name	Referring to upper margin of phyllodes which is twice toothed – one tooth rounded, the other acute.
Distribution	WA south-west in Avon and Stirling districts, e.g. Esperance, Ongerup, near Dowerin; on sand in shrubland on inland plains.
Habit	Extremely variable, much-branched shrub or under-shrub, prostrate to 1 m × 0·6–1·5 m; sometimes a dense, dome-shaped shrub to 3 m tall with many grey stems, hairless or with a scattered or dense covering of short hairs; branchlets occasionally ending in spines.
Foliage	Phyllodes variable, dull green, wedge-shaped (2–) 6–8 (–10) mm × 2–5 mm, often covered with short white hairs; main nerve near lower margin, terminating in a small acute point; upper margin rounded to form a blunt angle; stipules brown, persistent, occasionally developing into spines.
Flowers	Numerous cream, occasionally yellow balls 6–8 (–10) mm diameter, each of 8–15 or more flowers (sometimes flower-heads are slightly elongated) on hairless or hairy stalks 6–9 mm long, singly or in pairs in axils. Flowering August–November.
Pods	Dark brown, occasionally hairy, curved and twisted 2·5–5 cm × 2–4 mm, margins thickened, slightly rounded over and constricted between seeds.
Seeds	Brown, oval, flattened 3–4 mm × 2–2·5 mm, longitudinal in pod.
Identification	Shape and venation of phyllodes, twice-toothed upper margin; construction of usually cream flowers.
Comments	A hardy, long-flowering, long-lived species requiring a warm position with good drainage. It is growing successfully in Melbourne. Strikes readily from cuttings.

Acacia bidentata

22 *Acacia truncata*

(Burm. f.) Hort. ex Hoffmannsegg

Synonym	*A. cuneata* Benth.
Common name	None known.
Meaning of name	Referring to abruptly cut off top of phyllodes.
Distribution	WA south-west in coastal areas from Leeman, 250 km north of Perth, south to Bunbury; restricted almost entirely to shallow sand over limestone; frequently a common plant of the coastal heath vegetation.
Habit	Usually a dense, dome-shaped shrub 0·5−2·3 (−3) m × 1−1·5 m with smooth, light grey bark and upright branches; branchlets finely or prominently ribbed, hairless or hairy.
Foliage	Medium to dark green, hairless, ± wedge-shaped phyllodes 9−25 (−40) mm × 5−13 (−16) mm across top of wedge, straight or occasionally curved, very slightly wavy, yellowish margins; prominent mid-nerve, rarely nearer lower margin, ending in a firm, not sharp point; circular gland (rarely two) at top angle of wedge; stipules usually persistent.
Flowers	Light yellow balls 6−7 mm diameter, each of 7−16 flowers, on hairless, rather slender stalks 10−18 mm long, in a much reduced raceme of 1−2 flowers at top of stems; bracts persistent. Flowering June−September.
Pods	Dark grey-black, hard, brittle, usually hairless, curved or twisted 6·5−9 cm × 2−4 mm with thickened, yellowish margins, barely raised over and slightly constricted between seeds, tapering or ending abruptly at tip.
Seeds	Brown, shining, oblong 3−3·5 x 1·5−2 mm, longitudinal in pod; seed-stalk thread-like, tiny, abruptly thickened into a club-shaped, yellow aril.
Identification	Similar to *A. littorea*, but differs in its normally longer, narrower phyllodes, usually persistent stipules, earlier flowering season, more northerly occurrence, mostly on sand over limestone. The two species are closely related and further botanical study may change their status.
Comments	A small shrub for coastal planting; it is growing successfully in clay and in partial shade in Melbourne. Grown from cuttings and seeds. This plant was described originally as a fern. It has been reported as being possibly one of the first two plants collected by Europeans in Australia.

Acacia truncata

23 *Acacia phaeocalyx*

Maslin

Common name None known.

Meaning of name Referring to the dark brown calyx.

Distribution WA in Avon district between Wongan Hills, Kellerberrin and Tammin, occasional on sand over laterite in tall shrubland.

Habit Intricately branched, spreading or compact shrub 0·3–1 (–2) m tall with many round, finely ribbed, brown to grey branches; branchlets hairless, frequently covered with bloom. New growth bright red.

Foliage Green or grey-green, rigid, leathery, ± triangular, slightly wavy top margin 8–15 mm × 6–11 (–14) mm at widest point, with yellowish main nerve near lower margin, minor veins diverging, terminating in a long, very sharp point; margins slightly thickened, yellowish; sharply reduced at base; glands small, 1 or occasionally 2, on slight angle of top margin 6–9 mm from base; stipules spiny, recurved, persistent 2–4 mm long.

Flowers Large, loose, golden yellow balls 9 mm diameter, each of 4–8 flowers, on hairless stems 5–10 mm long which are normally hairy at the extreme base, singly or in pairs in upper axils; calyx dark brown, conspicuous, with broad triangular lobes. Flowering April–June; flowers have been collected in September.

Pods Dark red-brown, faintly veined longitudinally, hard, brittle, round 5–8 (–11) cm × 3–5 mm, tapered at both ends, curved into a half circle, not constricted between seeds. Immature pods red.

Seeds Mid-brown, 4–5 mm × 2–2·7 mm, longitudinal in pod; seed-stalk slightly curved and thickened into a cone-shaped aril.

Identification Sharp pointed, ± triangular phyllodes, spiny stipules, large flower heads of few flowers, dark brown calyx.

Comments Distinctive species; no information available regarding cultivation.

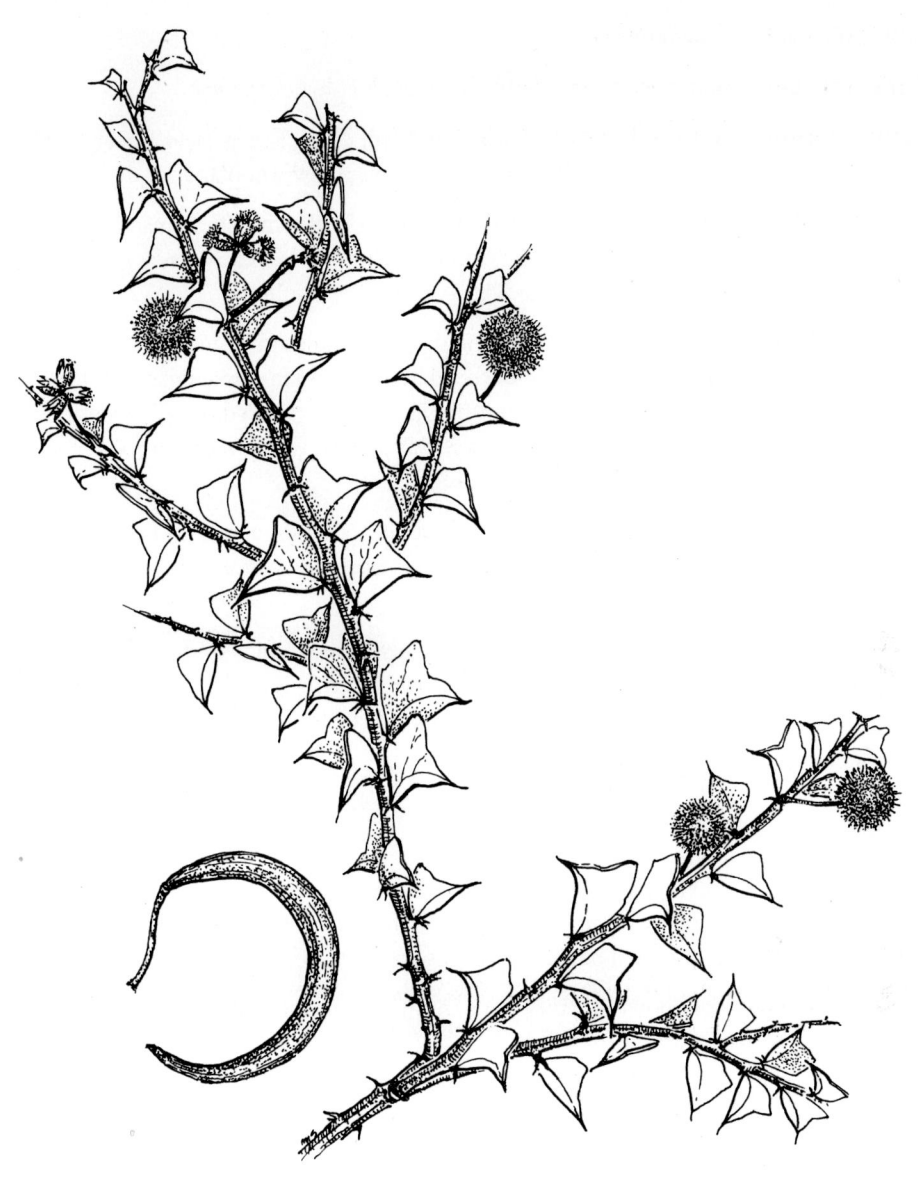

Acacia phaeocalyx

24 *Acacia dilatata*

Benth.

Common name	None known.
Meaning of name	Widened or spread out; referring to phyllode shape.
Distribution	WA south-west in Irwin and Avon districts, e.g. near Eneabba, Hill River and Badgingarra, often in open low woodland.
Habit	Spreading, much-branched shrub 45–50 cm tall; stems ± round, finely lined, with a covering of short, dense, white hairs. New growth densely hairy.
Foliage	Green, normally hairy, rigid, ± triangular phyllodes 1–2 cm long, as wide as long, with thickened margins, main nerve near straighter lower margin, minor nerves running towards rounded, wavy, upper margin; at tip drawn out into a very sharp, long, straight or curved point; gland obscure; stipules bristly or spiny.
Flowers	Mid-yellow balls 8–9 mm diameter, each of about 20 flowers, on usually hairy stalks 5–10 mm long, solitary in upper axils. Flowering November–February.
Pods	Dark reddish-brown, hairy, finely lined, round, slightly curved, *c.* 3–4 cm × 3–5 mm narrowed towards each end, not constricted between seeds. Pods often held in clusters.
Seeds	Dark brown, oblong, 5–6 mm × 2–5 mm, longitudinal in pod; seed-stalk short, thickened into a large, brown, cap-shaped aril at the end of seed.
Identification	Rigid, triangular phyllodes with thickened margins and nerves; stiff, slightly curved pods. Phyllode shape is somewhat like some forms of *A. truncata* but flower-heads are very different.
Comments	Not known in cultivation. Some plants were found to be heavily parasitised.

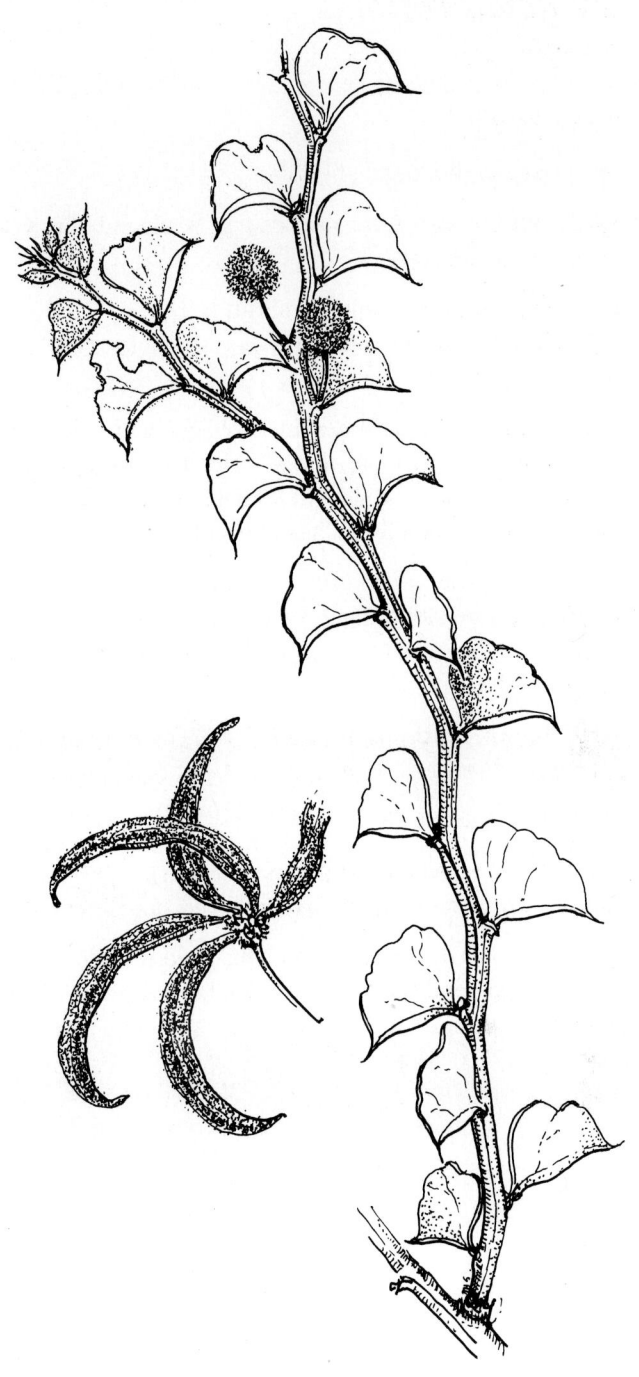

Acacia dilatata

25 *Acacia orbifolia*
Maiden

Group 4

Common name	None known.
Meaning of name	Referring to the round phyllodes.
Distribution	WA south-west in Avon district, e.g. Wongan Hills, Kellerberrin, Southern Cross areas.
Habit	Compact, dark green foliaged shrub 1–1·5 m × 2–3 m with grey branches; branchlets round, white scurfy-hairy. New growth often deep red.
Foliage	Deep green, shining, rather thick, obliquely rounded to round phyllodes 5–12 mm long and as wide; margins minutely hairy, top margin finely wavy, ending in a small, brown point; several fine nerves spreading from base; small gland near base.
Flowers	Masses of dense, cream balls 8–9 mm diameter, each of 20–30 flowers, on smooth or hairy stalks 10–25 mm long, singly or in pairs in upper axils; calyx cup-shaped with wavy edge. Flowering August–October.
Pods	Dark brown, brittle, tightly curled, *c.* 3 cm × 3 mm raised over and constricted between seeds.
Seeds	No information available.
Identification	Close to *A. bidentata*, but differs in its larger, rounded phyllodes, longer flower stalks, much larger flowers and different shaped calyx.
Comments	A most attractive small shrub with a good contrast of cream flowers and deep green phyllodes, which would be an interesting addition to the home garden.

Acacia orbifolia

26 *Acacia lycopodiifolia*

A. Cunn. ex Benth.

Group 5

Common name None known.

Meaning of name *Lycopodium* — a club moss; referring to the club-moss-like phyllodes.

Distribution WA north-west, Kimberley region into adjacent areas of NT, with a gap of about 240 km from Victoria River to eastern end of its range at Katherine Gorge; among sandstone ridges on dry, shallow, stony soils.

Habit Low growing, much-branched, spreading shrub to 1 m tall; branches covered with dense or sparse, sticky, white hairs; not prominently ribbed.

Foliage Phyllodes grey-green, very fine, ascending, straight or recurved 1·5–5 (–8) mm long, in regular whorls of (8–) 10–14 (–16), each with a faintly impressed nerve, abruptly narrowed into a fine, hairy, sticky point (0·2–) 0·4–1·6 (–4) mm long; stipules fine. Phyllodes on young plants normally are longer than those on flowering plants.

Flowers Bright yellow balls 5–9 mm diameter, each of (20–) 30–40 flowers, on stalks as long as and often much longer than phyllodes; bracteoles conspicuous, as long as or longer than buds. Flowering June–August, sometimes earlier.

Pods Flat, almost straight, hairless or sometimes hairy 3–5·5 cm × 4–6 mm with slightly thickened margins, raised over and sometimes constricted between seeds. Young pods covered with long, white hairs.

Seeds Black, oval, 4–6 mm × 2–3·5 mm, longitudinal in pod; a small aril at base.

Identification It has been confused with *A. adoxa*, but differs in its quite distinct, short, fine phyllodes with long, bristly points and its longitudinally placed seeds.

Comments An attractive small shrub which would be an asset in a rockery in the hotter regions.

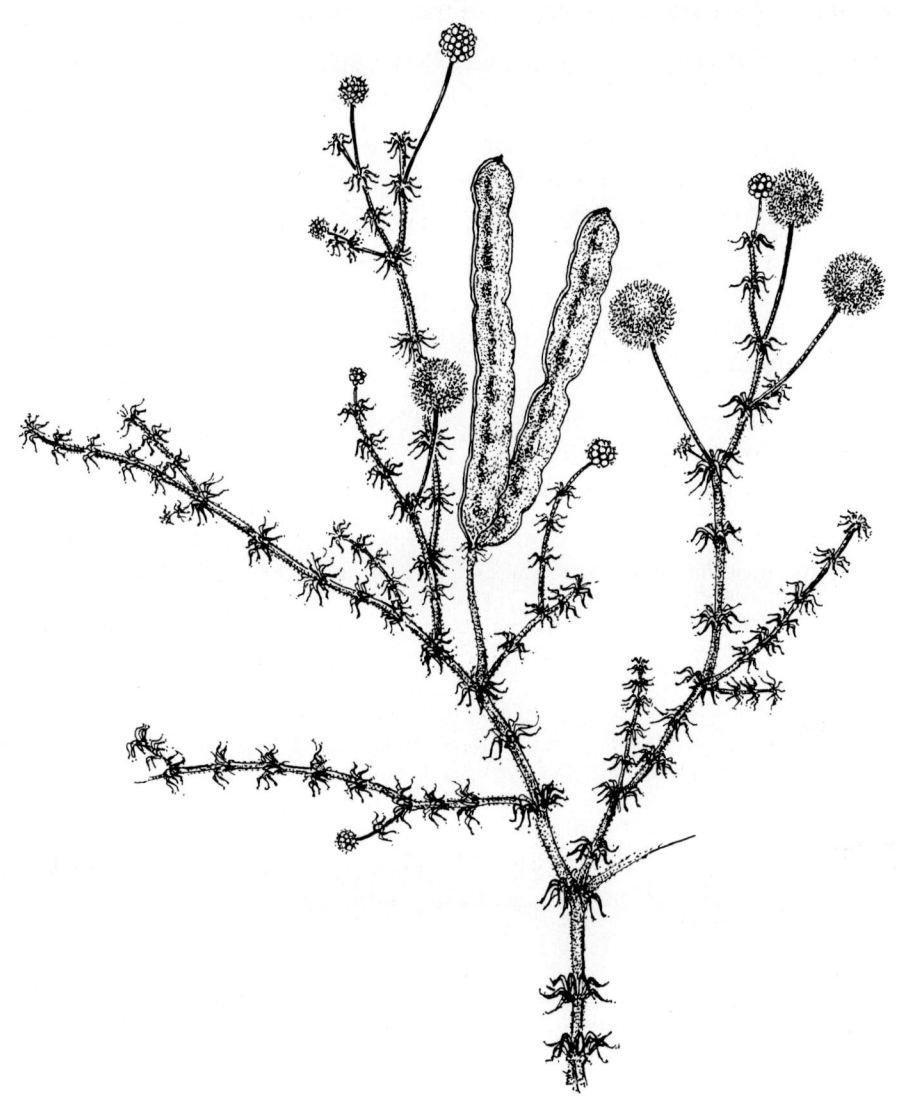

Acacia lycopodiifolia

27 *Acacia cometes*

C. Andrews

Common name	None known.
Meaning of name	Referring to covering of fine transparent hairs.
Distribution	WA in Eyre district, e.g. near Ravensthorpe to Esperance, in sandy or clay soil.
Habit	Prostrate, spreading, heath-like shrub 15–45 cm × 1·3–2 m with rough grey bark; branchlets densely hairy at tips, lower down sharply rough where phyllodes have fallen.
Foliage	Blue-green, numerous, crowded, erect, very fine phyllodes 8–12 (–15) mm × 0·5–0·75 mm, finely nerved and longitudinally wrinkled, tapering into an acute but not sharp point; stipules brown, persistent.
Flowers	Masses of dense, bright yellow balls 4–6 mm diameter, each of 20–25 flowers, on hairless stalks 5–10 mm long, usually solitary, crowded in the upper axils. Flowering September–October.
Pods	Dark brown, loosely curled or tightly coiled 4–5 cm × 1·5–2·5 mm, a little raised over seeds and slightly constricted between them.
Seeds	Brown-black, oblong, 2–3 mm × 1·5 mm, longitudinal in pod; seed-stalk fine, folded into a fleshy, frilled, cap-like aril.
Identification	Numerous, very fine, erect phyllodes crowded along branches; curled or tightly coiled pods. Its closest relative is *A. lachnophylla* which has longer phyllodes, longer flower-stalks and larger flower-heads.
Comments	Shrub suited to full sun or a little shade, in well-drained loamy, sandy or clay soils. It is considered long-lived in favourable conditions and is being grown more frequently in eastern states.

Acacia cometes

28 *Acacia gordonii*

(Tindale) Pedley

Synonym	*A. brunioides* A. Cunn. ex G. Don ssp. *gordonii* Tindale
Common name	None known.
Meaning of name	Named after Mr E. Gordon who first drew Dr Tindale's attention to the shrub.
Distribution	Rare species, confined almost entirely to the Blue Mountains, NSW, on sandstone. There is one record from Hornsby (a northern suburb of Sydney).
Habit	Heath-like, often straggling shrub 0·5–1·3 m tall with ± light brown branches densely covered with short, white hairs.
Foliage	Green, rough, linear, ± round phyllodes 10–15 mm long, covered with spreading, usually curved, white hairs, irregularly whorled around stem, ending in a fine curved tip.
Flowers	Perfumed, bright yellow balls 10–12 mm diameter, each of 20–30 flowers, on solitary, stout, hairy stalks 10–15 mm long. Flowering August–September.
Pods	Blue-black, oblong, flattened, 3·5–4 (–6) cm × 9–15 mm, with prominent reddish margins, little constricted between seeds.
Seeds	Black, oval, 3–4 × 2 mm, transverse in pod; seed-stalk short, ending in a cap-like aril.
Identification	Sometimes confused with *A. baueri* ssp. *aspera*, also a small shrub of the Blue Mountains. *A. gordonii* differs in its larger flower-heads with more numerous flowers, its blue-black pods and transverse seeds. It mainly differs from *A. brunioides* (a larger shrub which occurs in NSW and extends to south-eastern Qld) in its usually longer, thicker phyllodes, its larger flower-heads on longer stalks.
Comments	A beautiful small shrub which prefers a light soil, with some protection. It has been found to be very slow growing in Melbourne. This shrub has been raised to specific rank recently. Previously it was known as *A. brunioides* ssp. *gordonii* and it is possible that some botanists may not agree with the change of status.

Acacia gordonii

29 *Acacia minutifolia*

F. Muell.

Common name	None known.
Meaning of name	Referring to the very small phyllodes.
Distribution	Confined to NT, e.g. in Macdonnell Ranges, Valley of Glen of Palms, Mt Olga, and adjacent areas of WA.
Habit	A small rounded shrub 1 m $\times$ 1–1·5 m with sparse branches with a scattered covering of tangled hairs.
Foliage	Tiny, thick, smooth, blue-green phyllodes about 2·5 mm long, with wrinkled nerves and small recurved point; scattered but usually with 3 or 4 phyllodes together in a whorl around the stem; tiny stiff red-brown stipules at base.
Flowers	Bright yellow rather loose balls 8–9 mm diameter, each of 11–14 flowers, on stout hairy stalks about 10 mm long, singly at the tips. Flowering April–May.
Pods	Woody, sticky, flat, straight, 4–7·5 cm $\times$ 5–6 mm with wide margins, narrowing at base. Pods held erect on bush, curving open from the top.
Seeds	Small, brown, 3 mm $\times$ 1 mm, oblique in pod with a long, straight seed-stalk thickening only slightly at the end of the seed.
Identification	Arrangement and size of phyllodes, large flowers, shape and stickiness of pod and arrangement of seeds.
Comments	Grown from seeds; it would make an excellent subject for a dry inland garden with summer rainfall.

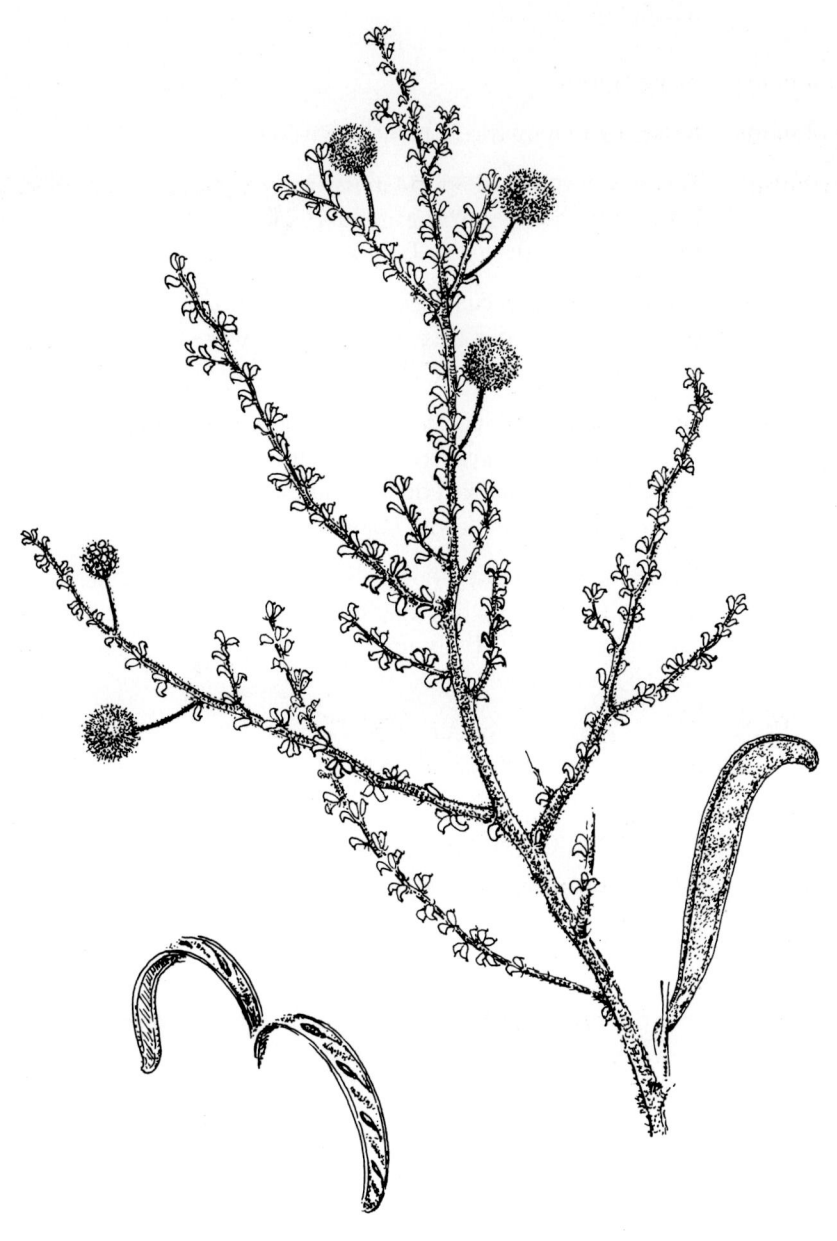

Acacia minutifolia

30 *Acacia fragilis*
Maiden & Blakely

Common name	None known.
Meaning of name	Referring to long fragile point of phyllode.
Distribution	WA south-west; widespread in Irwin, Eyre and Coolgardie districts in dry country between Wongan Hills to about 320 km north-east of Coolgardie, south to near Esperance.
Habit	Bushy, often mounded, needle-leafed shrub 1–3 m × 2–4 m with grey bark often fissured at base; branchlets grey or reddish, ± round. Young growth golden or silky hairy; mature phyllode tips, white hairy.
Foliage	Green, fine, round, needle-like phyllodes 2·5–7 (–8·5) cm × 1–1·5 mm, slightly curved, with 8 fine longitudinal nerves, tapering into a long, fine, hooked tip, normally carrying white, feathery hairs; gland, often obscure, about halfway along top margin.
Flowers	Masses of very bright yellow balls 5–9 mm diameter, each of 25–30 flowers, normally on fine, hairless stalks 2–5 mm long, singly or in pairs in upper axils. Buds often slightly elongated in shape. Flowering July–November.
Pods	Brown, thin, wavy or curled, 3–7 cm × 1–2 mm with nerve-like margins, alternately raised over and constricted between seeds.
Seeds	Brown, oval with indented centre, 3 mm × 1·5 mm, longitudinal in pod; seed-stalk thread-like at first, then abruptly thickened into an oblique, club-shaped aril at side of seed.
Identification	Similar to *A. assimilis* but differs in having darker green phyllodes with fewer nerves.
Comments	Free-flowering shrub which requires a well-drained position in full sun. Slow growth has been reported from Melbourne.

Acacia fragilis

31 *Acacia pilligaensis*

Maiden

Common name	None known.
Meaning of name	Referring to the Pilliga Scrub where the shrub occurs.
Distribution	NSW, restricted to the dry sandy areas of north-western slopes and plains, e.g. Pilliga Scrub north of Coonabarabran, and Goonoo Forest near Dubbo, and Qld.
Habit	Tall, much-branched, fine green-leafed shrub to 4 m tall with grey bark. Young branchlets angular, brown, covered with glandular hairs.
Foliage	Very fine, bright green, round but slightly flattened phyllodes 2–4 cm × 1–1·5 mm with central nerve, sometimes indistinct, tapering at tip into a small recurved blunt point, and at base into a tiny stalk; gland at base and tip of phyllode.
Flowers	Very bright yellow balls 6–8 mm diameter, each of 20–30 flowers, on short, sparsely or densely golden hairy stalks 2–5 mm long, singly in the leaf axils. Flowering September–October.
Pods	Brown, thin-textured, straight or slightly curved, 4–7 cm × 3–4 mm with nerve-like margins, rounded over and a little constricted between seeds.
Seeds	Black, oblong, 3·5–4 mm × 2–3 mm, longitudinal in pod; seed-stalk folded once or twice and thickened into a cup-like aril at end of seed.
Identification	Golden glandular hairs on flower-stalks and branchlets; shape and venation of phyllodes.
Comments	Suited to warm, dry, inland areas in full sun; requires a well-drained position. Grown from seeds.

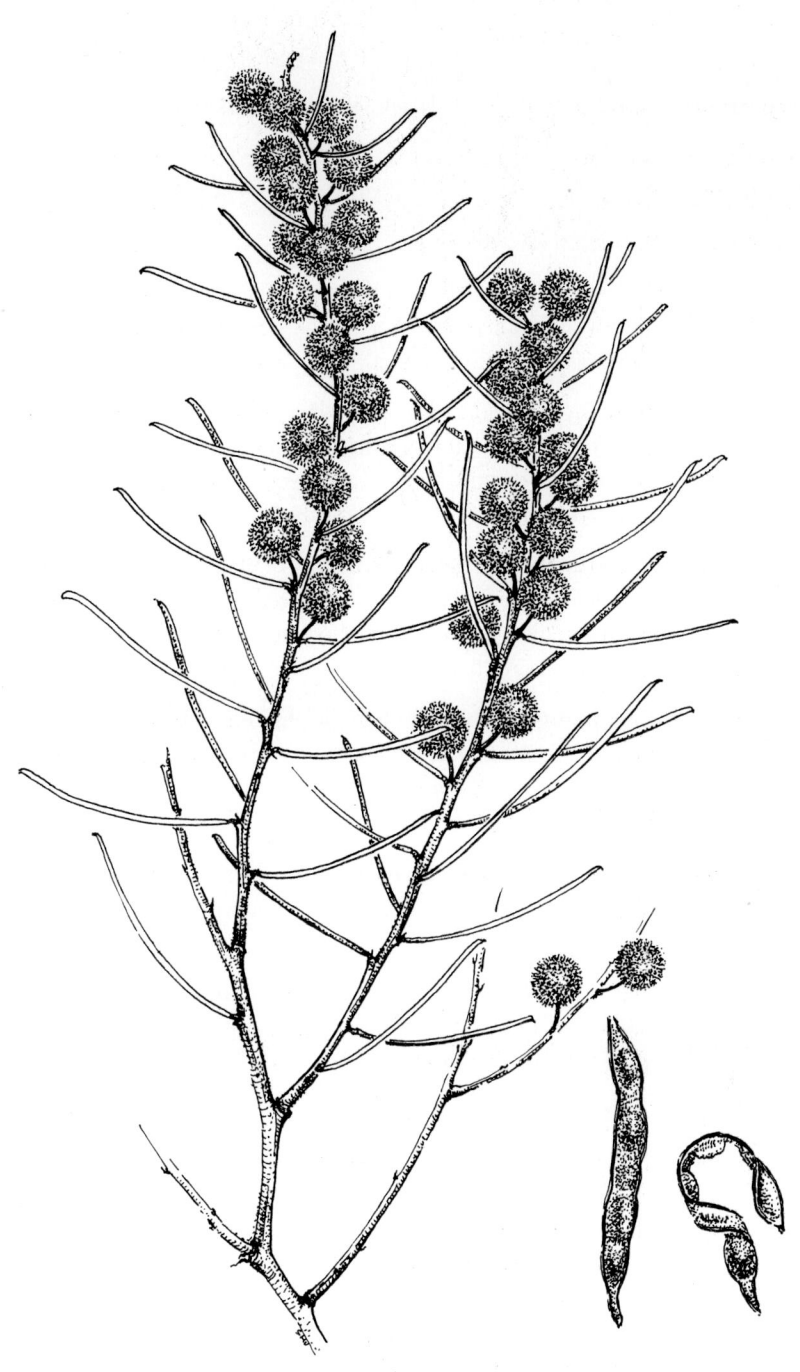

Acacia pilligaensis

32 *Acacia calamifolia*

Sweet ex Lindl.

Group 6

Synonym	*A. euthycarpa* J. M. Black; Whibley in *Acacias of South Australia*, 1980.
Common name	Wallowa
Meaning of name	With reed-like foliage.
Distribution	Widespread and common through a large part of the drier areas in open woodlands of north-west Vic., SA and NSW, usually on sandy loamy soils.
Habit	A tall or rounded shrub, smooth or slightly mealy, 2–4 m × 2–3 m but often taller and wider, with rather slender spreading branches and smooth brown-grey bark; branchlets angular at first soon becoming round. Young growth often mealy.
Foliage	Fine green to grey-green straight or curved phyllodes 3–12 cm × 1–5 mm, round, sometimes flattened with a longitudinal nerve on each side, always ending in a hooked point; small gland usually 5–10 mm from base.
Flowers	Masses of strongly perfumed, bright yellow balls 5–9 mm diameter, each of 30–40 flowers, usually 2–4 on short racemes, sometimes solitary or in pairs on smooth slender stalks 4–10 mm long. Flowering August–October, occasionally at other times.
Pods	Dark red-brown, often veined, curved, 6–15 cm × 3–6 mm raised over distantly placed seeds, evenly, slightly or much constricted between them.
Seeds	Black, oblong, 4–5 mm × 2–2·5 mm, longitudinal in pod; long seed-stalk often almost encircling seed, bending and folding back several times, ending in a fleshy aril.
Identification	Round or flattened phyllodes always with a hooked point.
Comments	Grown from cuttings or seeds; adaptable, hardy, ornamental shrub which is widely grown in southern states, including Tas.; it will tolerate some damp conditions.

Seeds are known food of Mallee Fowl (*Leiopa ocellata*) at Wyperfeld National Park, Vic.

A. calamifolia is a complex species which is to be revised. In the meantime, *A. nematophylla*, which was a synonym, will be recognised as a separate species. It is a shrub of coastal dunes with shorter phyllodes and bractless flower-stalks that do not form part of a raceme.

Acacia calamifolia

33 *Acacia extensa*
Lindl.

Group 6

Common name	Wiry Wattle
Meaning of name	Presumably referring to winged or spread out, flattened stems and phyllodes.
Distribution	WA south-west in Avon, Darling, Stirling and Warren districts, confined to the jarrah (*Eucalyptus marginata*) forests, especially along creek banks.
Habit	Slender, erect shrub 1–2 (–3) m tall with grey smooth bark; branches elongated, prominently angled, flattened, sometimes almost winged; branches and phyllodes often difficult to distinguish.
Foliage	Slender, stiff, erect or spreading, flattened, linear phyllodes, sometimes scaly rough (5–) 8–15 (–20) cm × 0·5–2 mm with prominent mid-nerve, ending in a small recurved tip.
Flowers	Numerous, bright yellow balls 8–10 mm diameter, each of 20 or more flowers, on hairy or hairless stalks 5–10 mm long, usually solitary, occasionally in irregular racemes. Flowering July–October.
Pods	Brown, thin-textured, straight or slightly curved, 8–10 cm × 4–5 mm with thickened margins, raised over and constricted between seeds. Immature pods reddish.
Seeds	Dark brown, shining, oblong, 4–5 mm × 2 mm, longitudinal in pod; seed-stalk folded and gradually thickened into a pale frilled aril.
Identification	Prominently ribbed, slender phyllodes and similarly ribbed, flattened, often almost winged stems.
Comments	Hardy, adaptable shrub which is grown widely in south-eastern states in many soil types in open to shaded positions and from well drained to wet conditions. It is frost tolerant. Grown from cuttings and seeds.

Acacia extensa

34 *Acacia juncifolia*
Benth.

Common name	Rush-leaf Wattle
Meaning of name	Referring to fine rush-like phyllodes.
Distribution	Widespread but not common on sandy soils in coastal or near coastal south-east Qld, e.g. near Charleville, Chinchilla to Blackdown Tableland, and the north and central coast and slopes of NSW.
Habit	Upright, very fine-leafed shrub 2–5 m tall with flaky brown bark and smooth, slender, rounded, brown branches.
Foliage	Erect or spreading green to blue-green, smooth, fine phyllodes 7·5–20 cm or longer × *c.* 1 mm, slightly flattened with one faint longitudinal nerve on each side ending in a very short straight or curved fine point; gland up to 10 mm from base, sometimes causing a slight bend.
Flowers	Many bright yellow balls 5–8 mm diameter, each of 20–25 flowers, on fine smooth stalks 4–10 mm long, singly or in pairs up stems. Flowering July–September.
Pods	Reddish-brown, smooth, thin textured, slightly wrinkled, 5–10 cm × 3–4 mm, evenly rounded over, lengthened and slightly narrowed between seeds, ending in a long thin point.
Seeds	Black, oval-round, lightly mottled, 3·5–4 mm × 2–3 mm, longitudinal in pod; straight, fine seed-stalk slightly thickened at end.
Identification	Very long, fine phyllodes with one nerve, flowers on simple stalks, mottled seeds.
Comments	Shrub suited to warmer inland or coastal areas of NSW and Qld. It is growing successfully in Rockhampton; plants are available from some native plant nurseries in Qld. Grown from seeds.

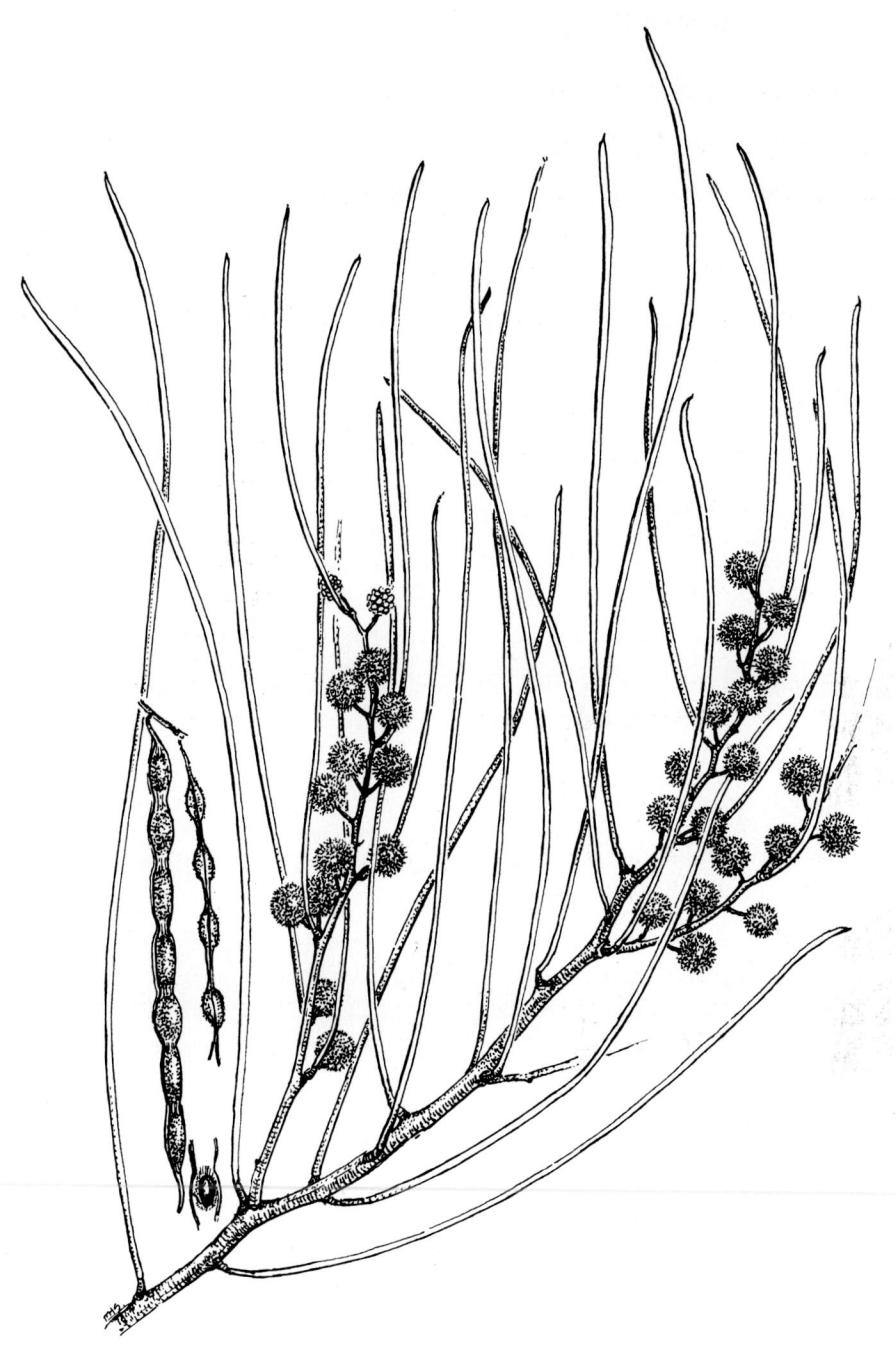

Acacia juncifolia

35 *Acacia camptoclada*
C. Andrews

Synonym	*A. porphyrochila* E. Pritzel
Common name	None known.
Meaning of name	Referring to flexible young shoots or branches.
Distribution	WA. Eyre and Coolgardie districts, e.g. south of Southern Cross to Balladonia area, in sand or clay.
Habit	Low, spreading shrub up to 1·6 m × 1–3 m with many slender branches from ground level; branchlets finely nerved, angular at first.
Foliage	Green to blue-green, stiff, oval-oblong phyllodes 0·6–1·5 cm × 2–3 (–4) mm crowded around stems; prominent mid-nerve, upper margin curved, with one or two well-spaced glands, ending in a fine, sharp, recurved point.
Flowers	Masses of bright yellow balls 5–8 mm diameter, each of 15–20 flowers, on fine, often reddish stalks up to 10 mm long, solitary, in pairs or clustered in upper axils. Flowering August–October.
Pods	Dark brown, curled into tight coils, *c.* 5 cm × 3–5 mm with slightly thickened margins, raised over and constricted between seeds. Immature pods glaucous, with purplish margins.
Seeds	Not known.
Identification	Crowded, stiff phyllodes, tip sharp, reflexed; position and number of glands; tightly coiled pods.
Comments	A decorative small shrub which would be suited to a well-drained position in full sun. Not known in cultivation.

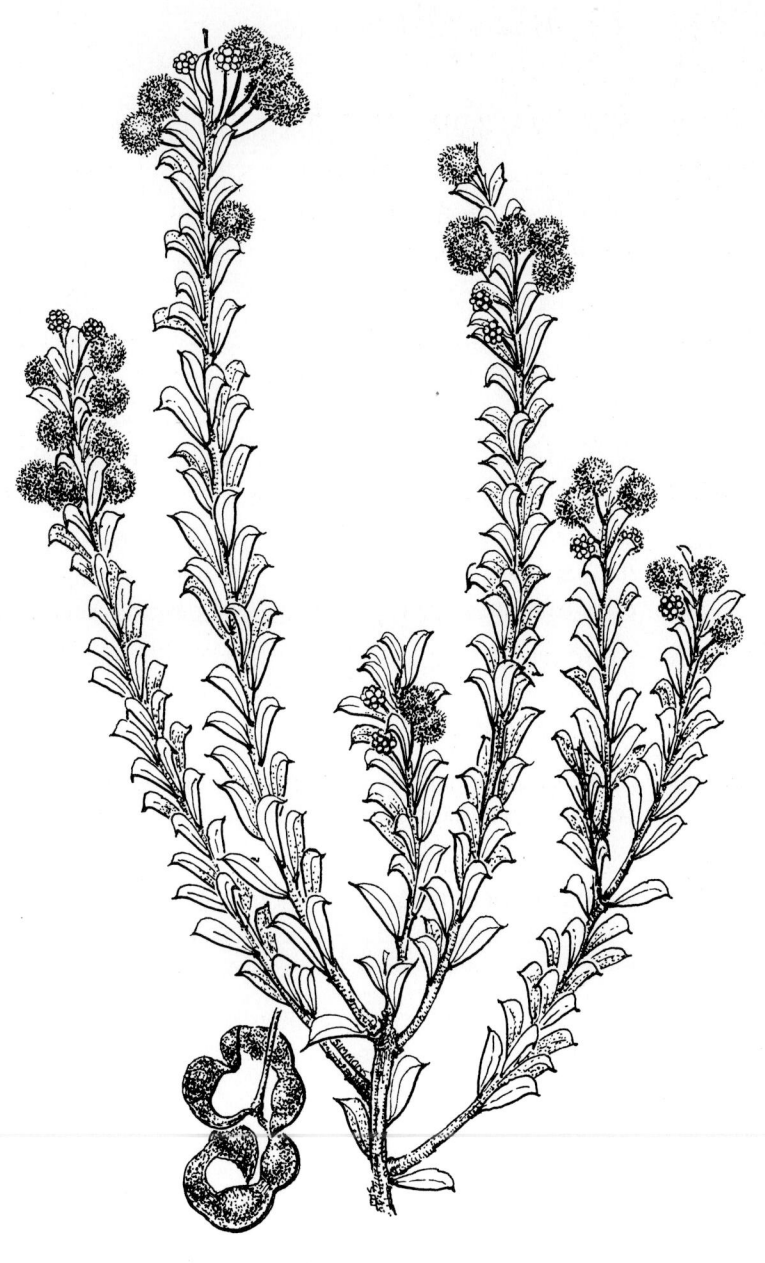

Acacia camptoclada

36 *Acacia spathulifolia*

Maslin

Synonym	*A. spathulata* F. Muell. ex Benth.
Common name	None known.
Meaning of name	Referring to spathulate or spoon-shaped phyllodes.
Distribution	WA in Irwin, Austin and Ashburton districts from near Jurien Bay, north to vicinity of North-West Cape; common on coastal limestone in sandy heaths.
Habit	Yellow-green, fleshy leafed, bushy shrub 1–3 m × 2–3 m with ± round branches, hairless or slightly hairy; branchlets round or angular, hairy.
Foliage	Yellow-green, ± fleshy, spathulate phyllodes 9–24 mm × 2–4 mm, faintly 1-nerved, margins thickened, ending in a tiny, blunt point or indented tip.
Flowers	Numerous, loose, bright yellow balls 6–8 mm diameter, each of 12–15 flowers, on hairless, stout stalks 5–10 mm long, singly or in pairs; calyx truncate or shortly lobed, hairy. Flowering August–October.
Pods	Fawn, veined, hard, straight or slightly curved, linear-oblong 2·5–4 cm × 3–6 mm, margins broad, tip blunt. Pods covered with short, dense hairs.
Seeds	Black, oval-oblong, 4–5 × 2 mm, longitudinal or slightly oblique in pod; seed-stalk short, thickened into a somewhat frilled, cup-shaped aril covering end of seed.
Identification	Distinguished by thick, fleshy, spathulate phyllodes; its very short-lobed calyx; flower-heads with few flowers. Some broad phyllode forms resemble *A. leptospermoides* ssp. *leptospermoides*.
Comments	An attractive shrub which has been grown successfully in Vic. It strikes readily from cuttings.

Acacia spathulifolia

37 *Acacia flexifolia*

A. Cunn. ex Benth.

Group 7

Common name	Bent-leaf Wattle
Meaning of name	With bent leaves.
Distribution	Most common in NSW in sandy soil on western slopes and plains in West Wyalong to Dubbo areas; restricted in Vic. to northern central areas near Bendigo; in Inglewood–Stanthorpe areas of southern Qld.
Habit	Small, slender, usually erect, grey-green or green-foliaged shrub 1–2 m tall with round stems, usually marked with raised hairy or resinous lines or ribs; young stems angular, scurfy or downy.
Foliage	Narrow, usually grey-green, smooth, $\pm$ erect bent phyllodes 0·9–2·5 (–5) cm $\times$ 1–2 mm with prominent nerve close to top margin, sometimes merging with it; usually broadening into an almost round tip or with a small point; narrowing at base into a small hairy stalk; an impressed gland near base where phyllode bends upwards.
Flowers	Very small, loose, perfumed, lemon-yellow balls 3–8 mm diameter, each of 4–8 flowers, on short, usually hairy stalks 2–4 mm long, singly or in pairs in leaf axils. Buds sometimes waxy white. Flowering April–September.
Pods	Thin, flat, narrow, light brown 3–7 cm $\times$ 2–2·5 mm with a few scattered hairs and slightly thickened margins, a little raised over and narrowed between seeds.
Seeds	Dark brown, oblong, 4·5–5 mm $\times$ 1·5–2 mm, longitudinal in pod; seed-stalk folded several times to form a basal aril.
Identification	Distinctly bent phyllodes with nerve near top margin, loose, few-flowered, lemon-yellow flower-heads.
Comments	Occasionally attains a greater height and a width of up to 4 m in cultivation. Frost resistant species; will withstand periods of dryness. Grown successfully in many situations from Tas. to Qld, it requires a well-drained sunny position; it may be pruned lightly after flowering. Reported to be fairly tolerant of salt-laden winds. Grown from cuttings and seeds.

Acacia flexifolia

38 *Acacia glandulicarpa*

Group 7

F. M. Reader

Common name	Hairy-pod Wattle
Meaning of name	Referring to glandular-hairy pods.
Distribution	Confined to several small areas, in open eucalypt scrubland in north-west mallee of Vic., in the Dimboola–Nhill region; the other on a rocky hillside in the Burra Gorge, Northern Lofty region of SA.
Habit	Low, often dense, wide-spreading green-leafed shrub 1–2 m × 2–3 m with grey bark and tough, hairy, rounded stems. Young growth crowded, shining bright green.
Foliage	Bright green, often wavy, smooth phyllodes, 0·5–2·5 cm × 2–5 mm, variable in shape from round to narrow-oblong, sometimes with a few scattered hairs, narrowing at base into a tiny stalk; mid-nerve ending in a small, soft, straight or recurved point; often with two glands, one at the tip and the other near the middle of the upper margin.
Flowers	Perfumed, bright yellow balls 5–6 mm diameter, each of 8–20 flowers, on smooth, hairless stalks 4–8 mm long, singly or in pairs in the upper leaf axils. Flowering July–October.
Pods	Brown, viscid, covered with stiff, glandular, long white hairs, straight or slightly curved, 1·5–3 cm × 3–5 mm with slightly thickened margins, little if at all constricted between seeds.
Seeds	Small, dull brown, oval, 3–4 mm × 2 mm, slightly oblique in pod; seed-stalk short, whitish, folded several times, thickening into a cap-like aril.
Identification	Wavy-edged, two-veined phyllodes, glandular-haired, sticky pods and low spreading habit.
Comments	Easily grown from cuttings or seeds; widely cultivated as far south as Tas. in a variety of soils including clay, in dry open or semi-shaded positions. It appears to be frost tolerant but does require good drainage.

Acacia glandulicarpa

39 *Acacia acinacea*

Lindl.

Common name	Gold-dust Wattle
Meaning of name	Curved, sword-like; referring to some phyllode shapes, but these are extremely variable.
Distribution	Widespread in Vic., except East Gippsland, SA and the southern riverine plains of NSW in annual rainfall areas of 350 mm.
Habit	A variable, much-branched, smooth, green-leafed shrub to 2 m tall with smooth grey branches often spreading 2–4 m wide; branchlets yellow-green, angular at tips, occasionally with scattered hairs. Often an upright shrub but occasionally prostrate.
Foliage	Bright green, often wavy, smooth phyllodes, 0·5–2·5 cm × 2–5 mm, variable in shape from round to narrow-oblong, sometimes with a few scattered hairs, narrowing at base into a tiny stalk; mid-nerve ending in a small, soft, straight or recurved point; often with two glands, one at the tip and the other near the middle of the upper margin.
Flowers	Masses of bright yellow balls 6–8 mm diameter, each of 12–20 flowers, on fine smooth stalks up to 1·5 cm long, usually singly but often up to four flowers together; persistent brown-tipped bracts often at base. Flowering August–November.
Pods	Crowded, ± straight, curled or spirally coiled, bright green pods drying light brown, 2–5 (–8) cm × 2·5–5 mm, slightly raised over and a little constricted between seeds.
Seeds	Black, small, oval, 4 mm × 2 mm, longitudinal in pod; short seed-stalk thickened into a fleshy club-shaped aril. Seeds mature in December–January.
Identification	Position of gland(s) on phyllodes, smooth flower-stalks and masses of curled or coiled pods. *A. rotundifolia* Hook. is considered a synonym.
Comments	Grows readily from cuttings and seeds. It is a hardy frost and drought resistant shrub. It is a recommended shrub for Alice Springs and thrives in a wide range of soil conditions in Tas. Like most acacias it prefers good drainage. Light to medium pruning after flowering is often necessary to maintain bushiness.

Changes, to be published shortly, are planned which will affect the classification of some forms of *A. acinacea*. *A. triquetra*, which was included with it, will be treated as a separate species and many of the specimens now recognised as *A. acinacea* in South Australia will be referred to *A. triquetra*.

A form of *A. acinacea* with larger phyllodes and smoother, less angular branchlets appears to grade into another variable species, *A. microcarpa*.

Acacia acinacea

40 *Acacia brachybotrya*

Benth.

Group 7

Synonym	*A. spillerana* J. E. Brown; Whibley in *Acacias of South Australia*, 1980.
Common name	Grey Mulga
Meaning of name	Bearing short racemes of flowers.
Distribution	Common, often on sandy rises through a wide area of NSW, SA and north-west Vic. in low rainfall areas.
Habit	Variable, rounded, spreading, grey-green shrub 1–3 m × 5–6 m with grey bark and smooth silvery-white hairy branches; branchlets at first slightly angular soon becoming round. Young growth hairy.
Foliage	Phyllodes thick, grey-green, smooth or hairy, obliquely-oval to roundish, 1–3·5 cm × 5–20 mm with a central nerve, faint penniveins and a blunt tip; gland usually near middle of top margin. Usually hairy phyllodes become smooth with age; this habit varies in different areas.
Flowers	Dense bright yellow balls 6–8 mm diameter, each of 20–35 flowers, solitary on slender hairy stalks 8–17 (–20) mm long or in groups of 2–5 in short racemes, often growing out into a leafy stalk. Flowering August–October.
Pods	Dark brown, veined, often warty, straight or curved 3–8 cm × 5–9 mm with thickened margins, raised over seeds and slightly constricted between them.
Seeds	Black, oval, 4–5 mm × 3–3·5 mm, longitudinal to oblique in pod; short seed-stalk thickened at end into a club-shaped aril.
Identification	Shape and size of phyllodes and pods; degree of hairiness and arrangement of flowers. Closely resembles *A. argyrophylla* Hook. which has golden, hairy new shoots, longer phyllodes always covered with silky hairs, usually larger flowers and pods.
Comments	Hardy species suitable for arid conditions; widely grown in NSW, Vic., SA and occasionally in Tas. At Wyperfeld National Park in north-western Vic. the seeds provide food for the Mallee Fowl (*Leipoa ocellata*).

Acacia brachybotrya

41 *Acacia paradoxa*
DC.

Synonym	*Acacia armata* R. Br.
Common name	Kangaroo Thorn or Hedge Acacia
Meaning of name	Contrary to the usual type, strange.
Distribution	Widely distributed through mainland states; rare in south-eastern Qld; absent from NT and introduced into Tas.
Habit	Variable, much-branched, bushy, green, prickly shrub 2–5 m tall, with hairy or occasionally smooth, ribbed angular branchlets; rigid, spreading, spiny stipules.
Foliage	Variable green, obliquely-oblong, usually wavy-edged, hairy or smooth phyllodes 1–3 cm × 2–8 mm with an almost central nerve, tapering into a slender tip; armed with straight, spreading, spiny stipules 5–12 mm long, at base of phyllodes; tiny gland near base.
Flowers	Large, very bright yellow balls 8–9 mm diameter, each of 30–40 flowers, on solitary, sparsely hairy or hairless stalks (8–) 10–13 mm long or longer, sometimes as long as phyllodes. Flowering August–October; June–August in WA.
Pods	Stalked linear, usually covered with long dense or scattered white hairs, occasionally hairless, 3–7 cm × 3·5–5 mm, straight or curved, with thickened margins, rounded over seeds, rarely constricted between them.
Seeds	Brown-black, oblong, 3–5 mm × 2 mm, longitudinal in pod; seed-stalk with 2–4 folds a little thickened under seed; aril small.
Identification	Sharp, long, spiny stipules; wavy-edged green phyllodes and large flowers.
Comments	Widely grown, moderately drought resistant and frost tolerant shrub. Useful for protective hedges and for nesting birds; in some neglected areas it has become a pest and is classified as a noxious plant in Vic. Will withstand some salt-laden winds if protected by dunes or trees.

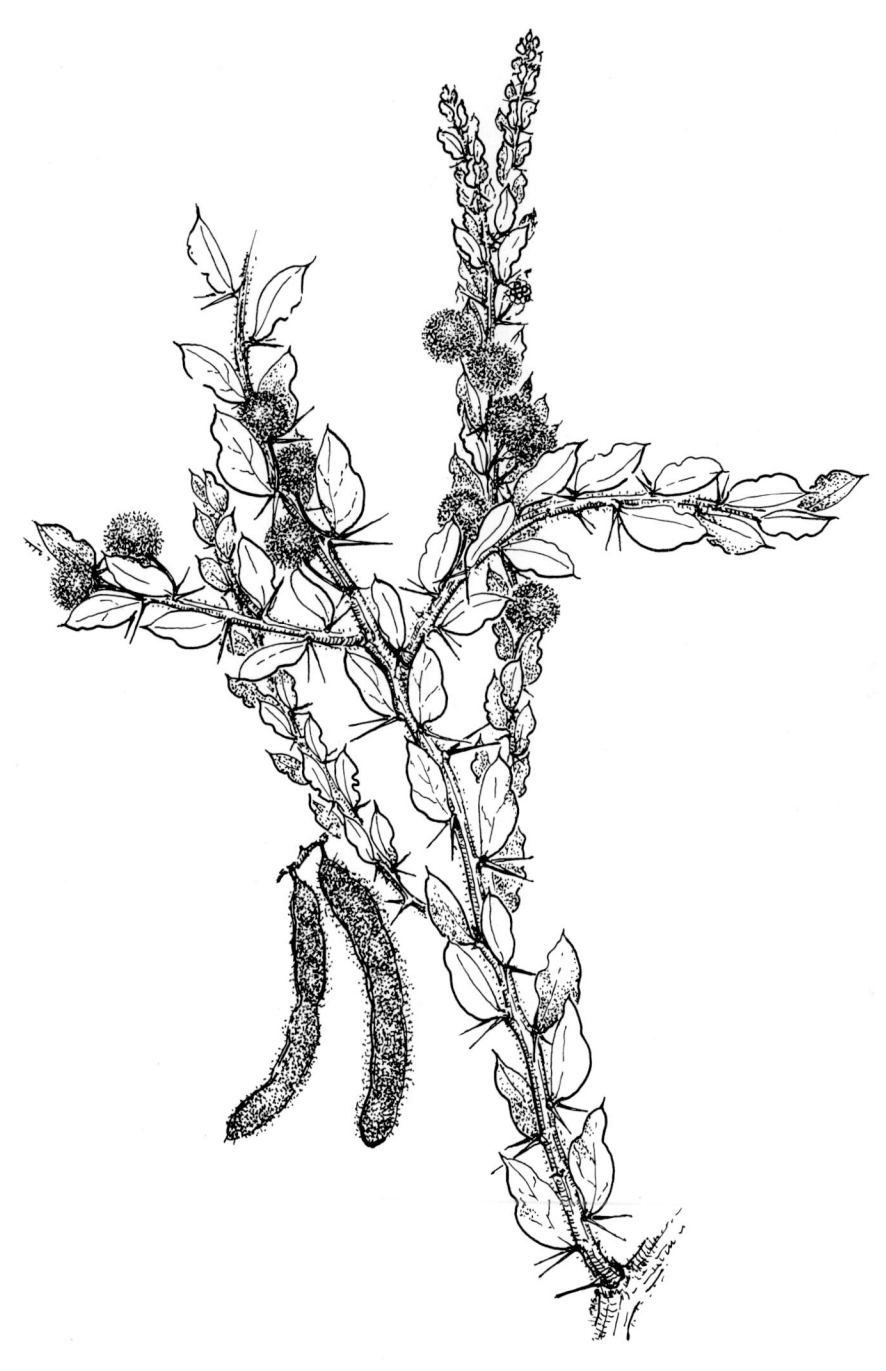

Acacia paradoxa

42 *Acacia aspera*
Lindl.

Group 7

Common name	Rough Wattle
Meaning of name	Rough to touch.
Distribution	Scattered through dry open forests and ridges of gold-bearing country in Vic., central western slopes and southern tablelands of NSW.
Habit	Resinous, rough leafed, erect or spreading shrub 1·2–3 m × 1–3 m with peeling grey bark on trunk; branches and branchlets roughly hairy and longitudinally lined or slightly ribbed. Young growth clothed with dense, sticky glandular hairs.
Foliage	Rough, dull green, somewhat undulating, linear-oblong phyllodes 0·8–3 cm × 2–7 mm, covered with short stiff glandular hairs; slightly thickened margins and central nerve, narrowing at tip into a blunt recurved point and at base into a tiny hairy stalk; stipules small and bristly.
Flowers	Buds resembling a burr caused by protruding bracts. Numerous dense pale or bright yellow flowers 5–8 mm diameter, each of 30–50 flowers, on rather thick hairy stalks 5–6 mm long, singly or in pairs. Flowering July–November.
Pods	Light brown, covered with very dense, often tangled, glandular hairs, linear, curved, 2–5 cm × 2–5 mm, rounded over and slightly constricted between seeds.
Seeds	Black, oval-oblong, 5 mm × 2–2·5 mm × 1·5 mm thick, longitudinal in pod; seed-stalk with short folds thickening at base into a cup-shaped aril.
Identification	Distinctive burr-like buds; glandular hairy phyllodes, pods and branchlets.
Comments	Widely grown shrub adaptable to light or heavy soils in dry open or semi-shaded positions in southern states; withstands frosts and tolerates very dry periods without water. Readily grown from cuttings or seeds.

Acacia aspera

43 *Acacia purpureapetala*

F. M. Bail.

Common name	None known.
Meaning of name	Referring to its purple petals.
Distribution	A very rare species restricted to a small area on gravelly rises in the Stannary Hills, west of Herberton in Qld.
Habit	Small, prostrate, hairy-leafed shrub with usually long, spreading, hairy branches up to 3 m in total width, or occasionally a small low shrub. New growth hairy, reddish-green, stems somewhat angular at tips. Flowers and fruits often on bush at the same time.
Foliage	Densely hairy at first, green to dull green, stiff, lance-shaped phyllodes 2–3 cm × 3–6 mm with raised margins, net-like veins and central nerve ending in a long oblique stiff point; tapering at base into a small felty stalk; two hairy stipules 2–2·5 mm long; gland near base.
Flowers	Dense mauve-pink balls 6–7 mm diameter, each of *c.* 20 flowers, on hairy stalks 8–12 mm long, singly along stem; usually not flowering in a mass but irregularly, a few flowers at a time, for a long period. Occasional plants are known to flower prolifically. Flowering June–July, sometimes as late as October–November.
Pods	Light, dull brown, glaucous, slightly wrinkled with a few scattered hairs 1·7–3 cm × 5–7 mm with nerve-like margins, rounded over and slightly constricted between few seeds.
Seeds	Dull black, oval, 4–5 mm × 2–4 mm, longitudinal in pod; threadlike seed-stalk ending in a small aril.
Identification	Mauve-pink flowers, small hairy phyllodes, usually long spreading prostrate branches.
Comments	Grown from cuttings and seeds; it is proving difficult to grow on, especially in colder southern states. It should be successful in dry, hot conditions in northern areas, and is known to have flowered in a Brisbane garden.

Acacia purpureapetala

44 *Acacia meisneri*

Lehm. ex Meisn.

Common name   None known.

Meaning of name   Commemorates C. F. Meisner (1800–1874), European professor of medicine and botany who collaborated with other botanists in the publication of descriptions of Australian plants.

Distribution   WA south-west in Avon district, e.g. near York on ridges, plains and roadsides on loam, clay or gravel.

Habit   Spreading, much-branched, fairly dense shrub 1·5–3 m × 2–4 m with smooth grey bark; young branchlets glaucous, angular, flattened. New growth light green and crowded.

Foliage   Blue-green, obliquely oblong-oval, somewhat wavy phyllodes 1·2–2·5 cm × 6–12 mm with prominent mid-nerve, faint minor veins, ending at tip in a small, soft, hooked point; narrowing considerably at base; gland small, about halfway along top margin.

Flowers   Numerous, perfumed, deep yellow balls 5–7 mm diameter, each of about 30 flowers, on hairless stalks 10–13 mm or longer, solitary or occasionally in branching racemes at tips of branches. Flowering irregularly, but mainly in summer.

Pods   In clusters, dark brown, thick, curved 7·5–12·5 cm × 5–9 mm, rounded over and evenly constricted between seeds.

Seeds   Dull black, oval, slightly flattened, 6–7 mm × 3–4 mm, longitudinal in pod; very long, much folded seed-stalk, the last fold almost encircling seed and returning, thickened at end.

Identification   Arrangement and shape of distinctive blue-green phyllodes, position of gland, characteristics of flower-heads.

Comments   A fast-growing shrub which is grown successfully in eastern states; it appears to be adaptable and frost tolerant. Grows readily from cuttings.

Acacia meisneri

45 *Acacia merrallii*

F. Muell.

Group 7

Common name	Merrall's Wattle
Meaning of name	Named after E. Merrall (1844–1913), an early collector of plants in WA and Vic.
Distribution	WA south-west in Darling, Stirling, Avon and Coolgardie districts; common from near west coast, through Norseman and further south, east to SA into Eyre and Yorke Peninsulas, on sandy or gravelly soils; favours clay soils in low-lying areas.
Habit	Variable, stiff, mound-like, occasionally erect shrub 0·3–2 m × 1–3 (–5) m; bark fissured at base; branchlets slightly angular, hoary or woolly-hairy, occasionally hairless. Young growth often deep red, usually woolly-hairy, becoming smooth.
Foliage	Variable, blue-green, stiff, flat, oval to almost round phyllodes 8–17 (–30) mm × 8–12 (–15) mm with yellowish central nerve and slightly wavy, thickened margins, ending abruptly in a curved or straight, sharp point; small gland towards middle of top margin occasionally absent.
Flowers	Numerous, perfumed, bright yellow balls 5–8 mm diameter, each of 20–35 flowers, on slender, sometimes red stalks 5–15 mm long, singly, in pairs or clusters in upper axils. Flowering August–October.
Pods	In clusters, dark brown, thin, much-curved, twisted or coiled 3–6 cm × 2–3 mm, raised over and a little constricted between seeds.
Seeds	Black, oval, *c.* 2·5 mm × 1·5 mm, longitudinal in pod; seed-stalk slender, with one or two folds ending in a yellow-orange cup-shaped aril covering nearly half the seed.
Identification	Flat, ± rounded phyllodes with thickened, yellowish margins, dark brown coiled pods, usually woolly-hairy young growth. A great deal of variation exists within this very widespread species and it may represent more than one species.
Comments	A free-flowering shrub which is being grown successfully in gardens. It requires an open sunny position with good drainage and it appears to be lime tolerant.

Acacia merrallii

46 *Acacia anceps*
DC

Common name	Two-edged Wattle
Meaning of name	Referring to two-edged flattened stems.
Distribution	On sand in coastal or near coastal low rainfall areas, restricted to Eyre Peninsula, nearby islands and southern Yorke Peninsula in SA and in WA.
Habit	Rigid, erect, smooth, blue-green shrub 1–3 m × 4–5 m with spreading grey branches; branchlets at first angular with prominent ribs developing into wings at base of phyllodes; stems often deep red.
Foliage	Phyllodes blue-green, very thick, stiff, obliquely oblong-oval, 1·5–5 cm × 1–3·5 cm, much thickened often wavy margins, mid-nerve and penniveins, abruptly rounded at tip, indented or with a small point; narrowed into a flattened stalk which at base usually continues as a wing running down stem; small gland not far from base.
Flowers	Large, very dense bright yellow balls 8–10 mm diameter, each of 60–130 flowers, on smooth stout stalks 10–25 mm long, singly at top of stem. Flowering mainly during summer.
Pods	Short-stalked, flat, woody, dull brown, covered with bloom, rough-surfaced, usually straight, 3–5 cm × 10–12 mm with wide thickened margins, little if at all constricted between seeds.
Seeds	Dark brown, oval, 4–5 mm × 3 mm, transverse in pod; seed-stalk much folded, half encircling seed, thickened into an oblique aril.
Identification	Thick phyllodes usually decurrent with flattened angular stem, position of gland, dense solitary flowers and woody flat pods.
Comments	Grown from seed; dense shrub useful as low shelter in sandy coastal areas of southern states.

Acacia anceps

47 *Acacia stricta*
(Andr.) Willd.

Common name	Hop Wattle
Meaning of name	Referring to stiff upright phyllodes and habit.
Distribution	Common and widespread in dry places in Vic., NSW, south-east SA, southern Qld (near Stanthorpe) and Tas. Often suckers and colonises dry hillsides and roadsides.
Habit	Variable, erect, tall, smooth, occasionally sticky, dull green shrub 1–5 m tall with angular ribbed branchlets. Often gives the appearance of being yellow-green.
Foliage	Dull, stiff, often yellow- or blue-green, smooth, narrow, elliptical phyllodes 3·5–13·5 cm × 3–10 (–17) mm often slightly twisted, with prominent mid-nerve, occasionally a second nerve, fine laterals, ending in a blunt short, curved or straight tip; narrowing at base into a smooth wrinkled stalk; a gland near base, if present.
Flowers	Pale to mid lemon-yellow balls 6–8 mm diameter, each of 20–30 flowers, on smooth, hairy or mealy stalks 2–8 mm long, usually in opposite pairs or clusters up the stem. Flowering August–October.
Pods	Light brown, sometimes sticky, thin, flat, straight, 5–10 cm × 2·5–5 mm with lighter, slightly thickened margins, little if at all constricted between seeds.
Seeds	Black, oval, 3–4 mm × 1·5–2 mm × 1–2 mm thick, longitudinal in pod; last fold of seed-stalk thickened into a cup-shaped aril.
Identification	Dull yellow-green upright phyllodes; usually paired, opposite, lemon-yellow flowers.
Comments	Very hardy, frost tolerant; grows quickly in cool shady or open dry positions; will tolerate some poor drainage. Grown from cuttings and seeds.

Acacia stricta

48 *Acacia salicina* Group 7

Lindl.

Common name	Broughton or Native Willow; Cooba
Meaning of name	Willow-like, referring to its habit.
Distribution	Widespread in dry inland of SA, Vic., NSW, NT and Qld, preferring moist areas near banks of creeks but also found in dryer areas in annual rainfall regions of 150–350 mm.
Habit	Usually a graceful, willowy, smooth, tall shrub or tree with a wide crown, up to 15 m tall with rough fissured bark and trunk up to 60 cm in diameter, pendulous branches and foliage; sometimes scurfy angular branchlets a little zig-zag at first. In some northern areas it is a shrub 1–3 m tall with shorter phyllodes; often suckering freely.
Foliage	Variable, smooth, pale green to blue-green, pendulous, brittle, thickish straight or curved phyllodes 4–17·5 cm × 4–25 (–32) mm with prominent central nerve and faint laterals, slightly wavy margins when wide, tapering into a fine, blunt, often recurved point and at base into a small, smooth stalk; a gland at varying distances from base and another occasionally at tip – sometimes one or other is absent. Phyllodes vary in size and shape on the same bush.
Flowers	Large, dense, pale yellow balls 10–12 mm diameter, each of 15–25 or more flowers, either singly on smooth stout stalks 5–15 mm long or in short racemes or panicles of 2–8 or more flowers, 1–6·5 cm long. Flowering during most months of the year; peak April–June.
Pods	Fawn to light brown glaucous, rounded, woody, usually straight, 3–12 cm × 6–10 (–12) mm with thickened margins, usually not constricted between seeds.
Seeds	Shiny, dark brown-black, roundish, 4·5–6 mm × 3·5–4·5 mm × 2 mm thick, longitudinal in pod; usually scarlet seed-stalk thickened and folded several times under seed.
Identification	Weeping willow habit, large pale ball flowers, usually straight, broad pods.
Comments	A fine long-lived tree for parks and gardens, widely grown in Townsville and a recommended tree for Alice Springs area. It is suitable for most soils and can withstand long periods without moisture. It is reported to tolerate saline conditions and some coastal exposure. Suspected of poisoning cattle in Qld, the bark was used by Aborigines to poison fish. The timber is close-grained, tough, dark brown.

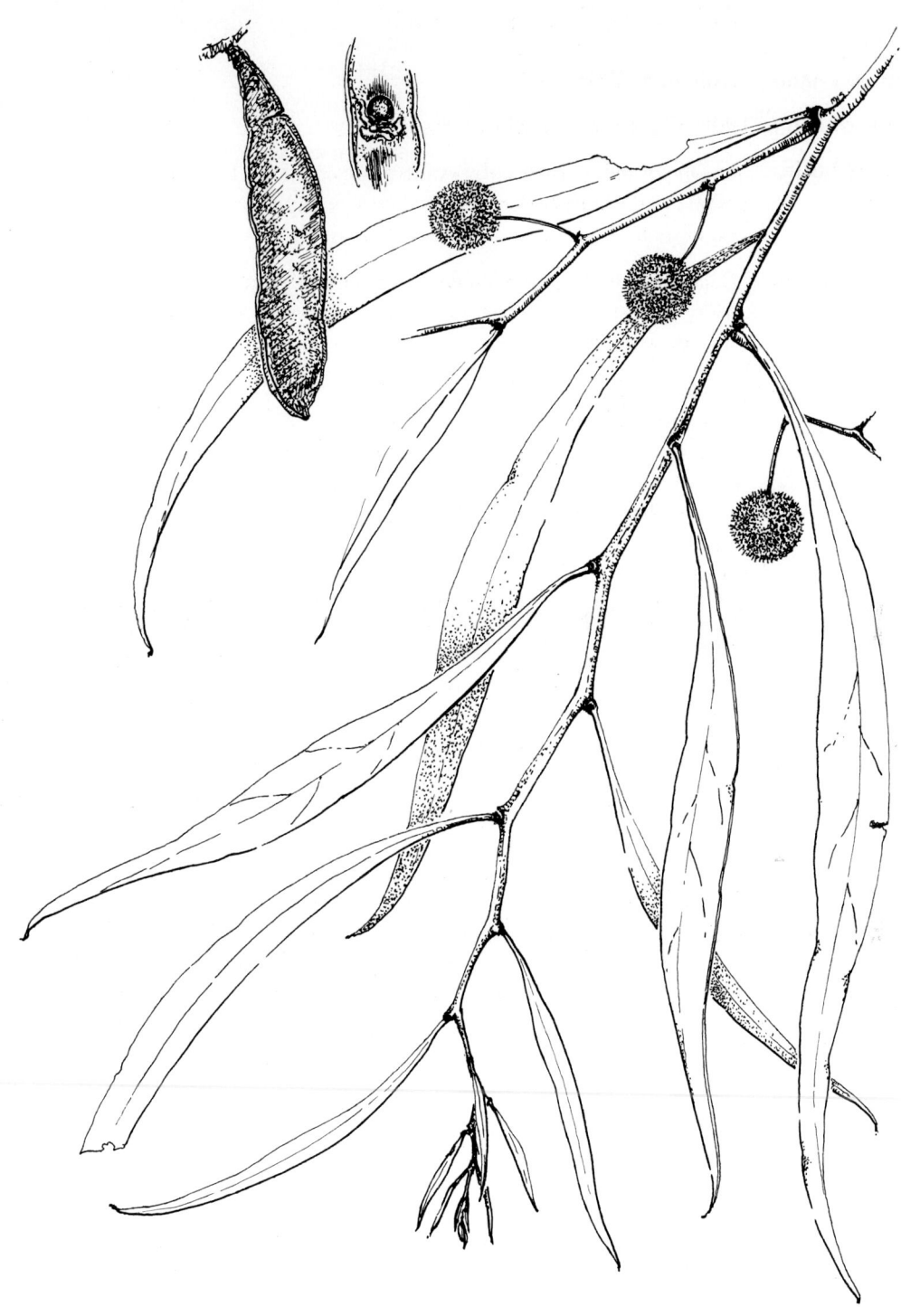

Acacia salicina

49 *Acacia cultriformis*

A. Cunn. ex G. Don

Common name	Knife-leaf Wattle
Meaning of name	Referring to blade-like phyllodes.
Distribution	In forests and on ridges of western slopes, tablelands and plains of NSW, e.g. in Pilliga Scrub near Coonabarabran, and into the Inglewood–Stanthorpe regions of southern Qld.
Habit	Bushy or spindly, smooth, blue-green shrub 2–4 m tall, sometimes as wide, with smooth, slender grey branches; branchlets smooth, angular ribbed, covered with bloom.
Foliage	Thick, smooth, usually crowded, obliquely oval or triangular-shaped blue-green phyllodes 1·2–3·5 cm × 7–14 mm at widest, with usually angled upper margin; mid-nerve nearer almost straight lower margin, penniveined; tapering into a short or long sharply curved point, and ending abruptly at base in a tiny narrowed stalk; a raised gland usually on or near angle of top margin.
Flowers	Numerous bright yellow balls 5–6 mm diameter, each of 13–40 flowers, on short smooth stalks (2·5–) 4–5 mm long, in racemes of 13–40 flowers, sometimes forming terminal panicles. One specimen collected south of Wyalong has small spike-like flower-heads 8 mm × 6 mm, 4–5 to a raceme. Flowering August–October.
Pods	Bunches of smooth, dark reddish-brown, thin-textured flat pods 5–7 cm × 5–7 mm with nerve-like margins, raised over and a little constricted between seeds. Immature pods covered with bloom.
Seeds	Black, oblong, 3–4 mm × 2–2·5 mm, longitudinal in pod; seed-stalk folded and thickened into a club-shaped aril.
Identification	Shape of phyllodes, position of gland and single nerve of phyllode.
Comments	Hardy, attractive shrub, frost and drought resistant; widely cultivated in a wide range of conditions from Tas. to Qld. It is growing in Atherton, Qld, but will not flower. Grown from cuttings and seeds.

Acacia cultriformis

50 *Acacia pravissima*

F. Muell.

Common name	Ovens Wattle
Meaning of name	Most irregular or crooked, referring to branching.
Distribution	Confined to banks of rivers and hills in highlands of north-eastern Vic., ACT and NSW in an annual rainfall area of 600 mm.
Habit	Large, bushy, dull-green leafed shrub or small tree 3–7 m × up to 7 m wide with smooth grey bark, often slightly pendulous branches and slender angular ribbed branchlets.
Foliage	Numerous, usually crowded, thick, usually dull green, roughly triangular or wedge-shaped phyllodes 6–18 mm × 4–14 mm with one nerve nearer the lower margin, often a lesser vein nearer the top; ending in a small point; narrowing at base into a small flattened stalk; small indented gland near base.
Flowers	Dense masses of small bright yellow balls 5–6 mm diameter, each of 6–9 flowers, on smooth stalks in racemes or long leafy sprays up to 10 cm long. Flowering August–November, but varies with altitude.
Pods	Flat, papery, smooth, 5–8 cm × 5–7 mm, slightly raised over seeds, occasionally narrowed a little between them. Pods often red when immature.
Seeds	Black, 3·5–5 mm × 2 mm, longitudinal in pod; seed-stalk thickened into a fleshy aril under seed with several small folds.
Identification	Somewhat similar to *A. cultriformis* but differing in shape and colour of phyllodes, position of gland and shape of pods.
Comments	Hardy shrub adaptable to most soil types in temperate Australia, grown widely in NSW, Vic. and Tas. Will withstand slight coastal exposure and is frost tolerant; a useful species for roadside plantings and for windbreaks. Can be pruned after flowering. A prostrate variety is available from some native plant nurseries. Grown from cuttings and seeds.

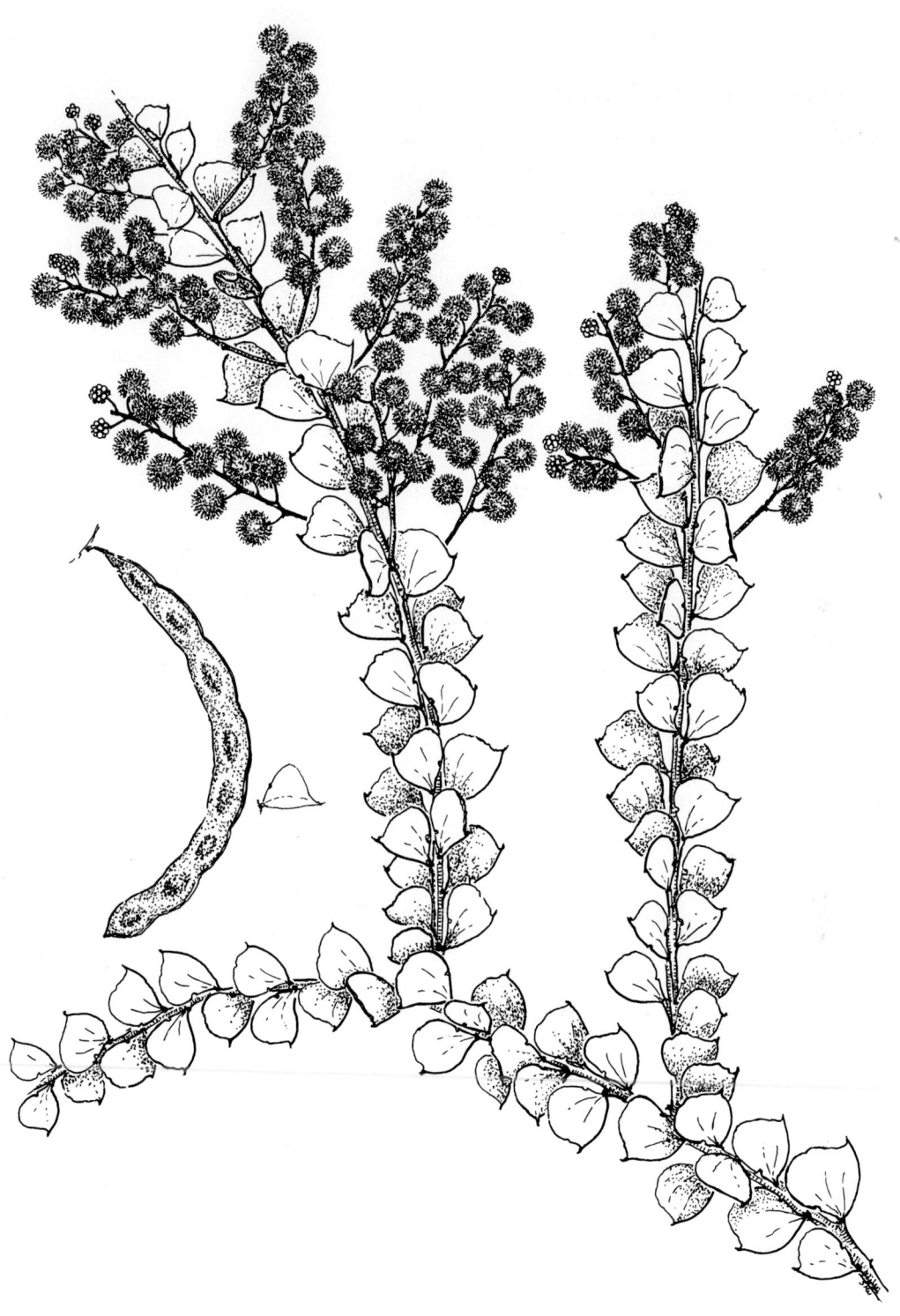

Acacia pravissima

51 *Acacia leichhardtii*
Benth.

Group 8

Common name   Leichhardt's Wattle

Meaning of name   Named for F. W. L. Leichhardt (1813–1848), the early explorer who first discovered the shrub.

Distribution   Confined mostly to rocky, dry sandstone areas of south-east Qld, usually on slopes and ridges, e.g. at Isla Gorge, Blackdown Tableland and forests near Chinchilla.

Habit   Erect, spreading, hairy-leafed shrub 1–3 m tall × 2 m or more wide, often with spotted grey bark and slender widely arching round branches covered with crowded, long, white hairs. New growth densely hairy.

Foliage   Usually green, hairy, straight or curved lance-shaped phyllodes reflexed down stem, 2–3 cm × 3–6 mm with near central nerve and nerve-like hairy margins ending in a fine, slightly curved, hairy point; phyllode narrows abruptly at base into a short hairy stalk; gland, if present, about halfway along top margin.

Flowers   Dense, very bright yellow balls 6–7 mm diameter, each of 25–40 flowers, on short, usually hairy stalks 3–8 mm long, in slender, spreading, hairy racemes up to 7 cm long. Flowering April–August.

Pods   Flat, curved, dark brown, made grey by long dense hairs, 5–10 cm × 8–10 mm with slightly thickened margins, rounded over and little or greatly constricted between seeds.

Seeds   Black, oval, 5–6 mm × 3–4·5 mm, longitudinal in pod; fine seed-stalk doubling back halfway around seed and then encircling it, ending in an elongated club-shaped aril.

Identification   Reflexed phyllodes, covering of long hairs, long slender racemes of very bright yellow flowers, densely hairy pods and seeds encircled by seed-stalk.

Comments   Grows successfully in Rockhampton, but needs very good drainage and is considered harder to grow there than many other acacias. Grown from seeds.

Acacia leichhardtii

52 *Acacia vestita*

Ker

Common name	Hairy Wattle, Weeping Boree
Meaning of name	Covered with hairs.
Distribution	Dry rocky areas mainly in central and southern tablelands and slopes of NSW, e.g. Mt Arthur near Wellington.
Habit	Weeping, soft blue-green foliaged shrub 2·5 × 4 m × 2–3 m with arching, densely hairy, round branchlets. New growth light green, densely hairy.
Foliage	Crowded, softly hairy grey-green, obliquely elliptical, sometimes undulating phyllodes 1–2 cm × 5–10 mm, with hairy nerve-like margins and near central nerve, narrowing sharply at tip into a fine, hairy, hooked, soft point, and at base into a tiny hairy stalk; no gland visible.
Flowers	Masses of bright yellow balls 6–7 mm diameter, each of 12–18 flowers, on slightly hairy stalks 4–5 mm long, in terminal racemes of 10–20 flowers, 1·5–5 cm long, often forming leafy panicles. Flowering August–October.
Pods	Flat, straight, bluish, smooth 3·5–7·5 (–10) cm × 9–12 (–15) mm rounded over and irregularly slightly constricted between seeds.
Seeds	Black, oval-oblong, 5–6 mm × 3·5–4 mm, longitudinal in pod; last fold of seed-stalk thickened into a long club-shaped aril almost as long as the seed, several fine folds below it.
Identification	Weeping hairy branchlets and small soft phyllodes; long racemes of bright yellow flowers and flat wide pods.
Comments	Versatile, hardy shrub grown widely from Qld to Tas.; drought and frost tolerant. Pruning after flowering is recommended to retain shape; it is reluctant to flower in Atherton, Qld. Grown from cuttings and seeds.

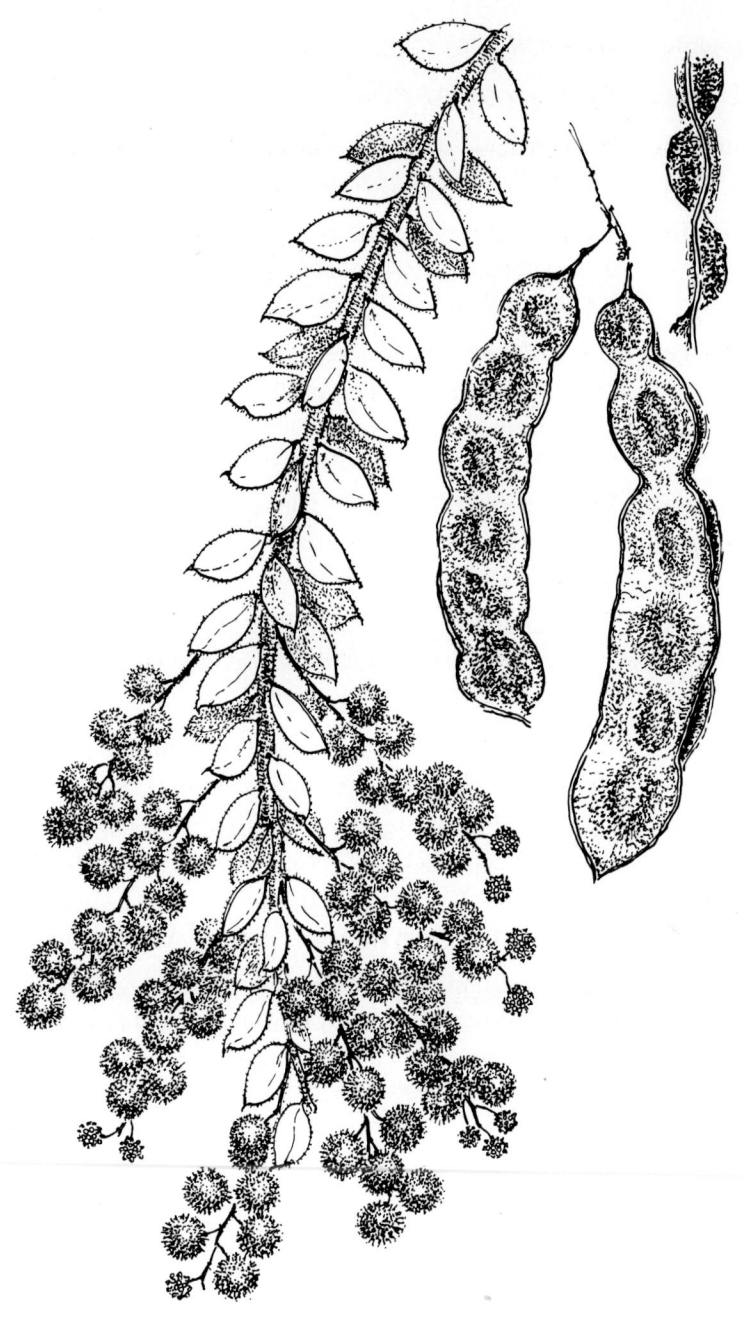

Acacia vestita

53 *Acacia buxifolia*

A. Cunn.

Group 8

Common name	Box-leaf Wattle
Meaning of name	With phyllodes like *Buxus*, the European box.
Distribution	Widely ranging species common in dry woodlands and highlands of Vic., ACT, NSW and south-eastern Qld about as far north as Tambo.
Habit	Variable, slender, smooth, blue-green shrub 2–3 m tall with usually erect, slightly angular, smooth branchlets, soon becoming round; often reddish in colour.
Foliage	Usually blue-green, obliquely elliptical phyllodes 1·3–3·5 cm × 3–11 mm, sometimes with a dull bloom; with a central nerve and narrowing at each end – at tip into a blunt or acute soft point; small gland up to halfway along top margin (sometimes absent).
Flowers	Masses of small, bright yellow loose balls 5–7 mm diameter, each of 8–20 flowers, on smooth stalks 3–6 (–8) mm long, in racemes 3–5 cm long. Flowering August–October.
Pods	Flat, narrow, straight or slightly curved, 6–10 cm × 7–9 mm coated with a little bloom, slightly thickened margins, rounded over seeds and slightly constricted between them.
Seeds	Black, oblong, 4–5 mm × 3 mm, longitudinal in pod; seed-stalk thickened into a short club-shaped aril.
Identification	Similar to *A. decora* Reichb. but differs in having smooth flower-stalks and racemes, and one gland. Subspecies *pubiflora* Pedley in *Austrobaileya*, 1, 3, 1979, is the more common subspecies in Qld; it has more elongate phyllodes and is found on sandstone.
Comments	Grown from cuttings and seeds, considered moderately frost and drought resistant. It is widely grown from Qld to Tas.
	A. furfuracea and *A. lunata* are two species which have been included under *A. buxifolia*, but will probably be shown to be distinct.

Acacia buxifolia

54 *Acacia decora*

Reichb.

Common name — Western Silver or Showy Wattle

Meaning of name — Pretty, graceful in appearance.

Distribution — Widespread and common on slightly elevated stony ridges and in sandy soils of western plains and tablelands of NSW, extending as far north as the Cooktown–Laura region of Qld; confined in Vic. to the Warby Range.

Habit — Variable, blue-green leafed, densely-flowered shrub 1–2 m tall, or a small tree 4–5 m tall, sometimes spreading to 4 m across, with smooth grey bark; branchlets angular, covered with appressed hairs or hairless.

Foliage — Variable, but basically lance-shaped, rather thick blue-green phyllodes 2–4 (–6) cm × (1·5–) 3–8 (–10) mm with prominent central nerve and minor branching veins; rounded or pointed at tip ending with a small point; tapering towards base into a small yellowish hairy stalk; gland near or some distance from base, a second gland sometimes present.

Flowers — Masses of perfumed bright yellow balls 7–8 mm diameter of 15–30 flowers, on stoutish golden-hairy stalks 3–6 mm long, in racemes usually longer than the phyllodes. Flowering June–September.

Pods — Brown, sometimes covered with bloom, flat, straight or curved, up to 11 cm × 4–9 mm, a little raised over seeds and usually constricted between them.

Seeds — Black, 4–6 mm × 2–3 mm longitudinal in pod; last fold of seed-stalk thickened into a club-shaped aril; lower folds very small.

Identification — Similar to *A. buxifolia* but differing in its terminal racemes of flowers usually exceeding the phyllodes, and stouter golden-hairy flower-stalks.

Comments — Hardy, drought resistant and moderately frost tolerant species, commonly grown in many areas in southern states. It grows best on sandy or stony soils in dry inland areas. Grown from cuttings and seeds.

Acacia decora

55 *Acacia prominens*
A. Cunn. ex Don

Common name	Gosford or Golden Rain Wattle; Grey Sally
Meaning of name	'Jutting out', referring to conspicuous gland on phyllode.
Distribution	Confined to creek banks and forests of coastal areas around Sydney and Gosford district, NSW.
Habit	Variable, grey-green leafed shrub or small tree 5–9 m tall, sometimes 20–25 m tall in more fertile soils in higher rainfall areas, with smooth grey bark and round branches; branchlets angular, usually covered with bloom. Young growth covered with bloom and often pinkish-grey in colour.
Foliage	Variable in shape, smooth, grey-green, lance-shaped, slightly curved phyllodes 2–6 cm × 5–10 mm with central nerve and minor penniveins; rounded tip ending in a small upturned hook; narrowing at base into a small wrinkled stalk; prominent raised gland on upper margin at variable distances from base.
Flowers	Masses of perfumed, small, bright lemon-yellow balls 4–5 mm diameter, each of 8–10 (–16) flowers, on smooth stalks, in slender racemes often exceeding phyllodes, 3–7 cm long. Flowering August–September.
Pods	Light brown, straight, flat, bluish with wrinkled surface 3–8 cm × 9–12 mm with thickened margins, a little constricted between seeds.
Seeds	Black, 5 mm × 3–4 mm, longitudinal, occasionally oblique in pod; seed-stalk in tiny folds thickened into a fleshy club-shaped aril.
Identification	Raised, round gland on top margin; few-flowered heads and long racemes of flowers.
Comments	Hardy, frost tolerant species; grows quickly in a wide variety of soils in fairly sheltered positions in an area of adequate rainfall. Useful as an ornamental tree and for planting in windbreaks.

Acacia prominens

56 *Acacia kettlewelliae*

Maiden

Group 8

Common name	Buffalo Wattle
Meaning of name	Named for Mrs Kettlewell, an original member of the Wattle Day League in NSW.
Distribution	Confined to gullies, hillsides and mountains of north-eastern Vic., around Mt Buffalo and Snowy River, and in adjacent regions of NSW; often growing with *Daviesea corymbosa* at elevations of 700–1000 m and more.
Habit	Variable, erect, blue-green leafed shrub or small tree up to 7–8 m tall with smooth grey-brown bark; branchlets brown, flattened acutely angular, soon becoming round but retaining slight longitudinal ridges. New growth bright light green.
Foliage	Blue-green lance-shaped phyllodes, greatly varying in size and shape, 3–10 cm × 3–12 mm with lighter coloured margins, central nerve ending in a small straight or curved blunt tip; narrowing at base into a smooth wrinkled stalk; usually one large indented gland not far from base. Broader phyllodes may have up to 3 glands, but this does not occur consistently.
Flowers	Masses of small bright yellow balls 4–6 mm diameter, each of 8–11 flowers, on smooth stalks 2–4 mm long, in dense racemes 4–6 cm long. Flowering September–November.
Pods	Flat, broad, purplish-red with bloom, smooth, thin-textured pods 3–10 cm × 7–12 mm raised alternately over seeds, slightly and irregularly narrowed between them.
Seeds	Black, oval, 5 mm × 2 mm, longitudinal or occasionally transverse in pod; seed-stalk short, thread-like, ending in a flattened white aril.
Identification	Resembles *A. prominens* but differs in number and placement of glands, calyx and petal characteristics, shorter and denser racemes.
Comments	Suitable shrub for cool temperate regions; frost tolerant. Grown from seeds.

Acacia kettlewelliae

57 *Acacia quornensis*

J. M. Black

Group 8

Common name	Quorn Wattle
Meaning of name	Named for Quorn in SA where it is found.
Distribution	Confined to hillsides near Quorn, northern Flinders Range, SA.
Habit	Upright, smooth, usually bushy shrub 2 m or more tall with smooth branchlets, at first angular.
Foliage	Pale or blue-green, smooth, lance-shaped phyllodes 2–5 cm × 4–5 (–9) mm with pale nerve-like margins and ± central nerve, tapering at tip into a small usually curved point and at base into a short stalk; small gland some distance from base.
Flowers	Pale lemon-yellow balls 6–7 mm diameter, each of 8–15 flowers, on smooth slender stalks 6–8 mm long in axillary racemes of 3–6 flowers, shorter than phyllodes; smooth, lobed cup-shaped calyx. Flowering September–December.
Pods	Rusty-brown, flattish, 5–13 cm × 7–10 mm with nerve-like margins, slightly constricted evenly between seeds and raised over them.
Seeds	Black, oval, 6–7 mm long, longitudinal in pod; white seed-stalk encircling seed in a double fold ending in a small aril.
Identification	Similar to *A. retinodes* var. *uncifolia* but differs in shorter, smaller phyllodes, fewer flowers in each head and smooth lobed calyx.
Comments	Grown from cuttings and seeds. It has been successfully grown in Melbourne; requires a warm well-drained position. Listed as an 'endangered' species in SA.

Acacia quornensis

58 *Acacia myrtifolia*
(Sm.) Willd.

Group 8

Common name	Myrtle or Red-stem Wattle
Meaning of name	With leaves like the European myrtle.
Distribution	Widespread and common in coastal heathlands, woodlands and slopes in temperate Australia from Canungra and Beerwah in Qld, south to Tas. and west to WA.
Habit	Variable, smooth, green-leafed often red-stemmed shrub, prostrate to 3 m tall with smooth acutely angular branchlets. Young growth and immature pods often brilliant red.
Foliage	Thick, smooth, variable in shape, basically obliquely lance-shaped phyllodes 2·5–8 cm × 5–20 (–30) mm, or in long, narrow-leafed variety 10–14 cm × 5–10 mm; erect, with prominent central nerve and penniveins; margins thickened yellow or red, ending in a sharp or blunt point; narrowing at base into a tiny stalk; small circular gland, sometimes prominent, 5–16 (–20) mm from base.
Flowers	Loose, large, cream to bright lemon balls 8–10 mm diameter, each of 2–8 flowers, on smooth, stout, erect stalks 3–10 mm long, usually in axillary racemes. Flowering May–August, sometimes later.
Pods	Dark brown, linear, curved, woody, flattened, somewhat twisted when mature, 2·5–11 cm × 2·5–4 mm with lighter coloured very thick margins often ending in a hard hooked point, a little irregularly narrowed between seeds.
Seeds	Shining, brown, oblong 3–4·5 mm × 1·5–2·2 mm, longitudinal in pod; seed-stalk very short, thickened into a fleshy cup-shaped aril.
Identification	Small number of flowers in each head, position of gland, woody pods.
Comments	Widely grown, long-lived hardy shrub; frost resistant, adaptable to a wide range of conditions. Pruning is sometimes necessary to retain shape and to encourage bright red new growth. Grown from seeds and cuttings. Narrow-leafed form is found in south-west WA and SA.

Acacia myrtifolia

59 *Acacia pataczekii*

Morris

Common name	Wally's Wattle
Meaning of name	Named for Wally Pataczek, the forester who first drew attention to the shrub.
Distribution	Confined entirely to several small areas in north-eastern Tas. highlands at Tower Hill and Roses Tier. It is found in a mixed eucalypt forest at an elevation of 1400 m.
Habit	Smooth blue-green shrub or small tree to 6 m tall with smooth grey bark and branches covered with whitish bloom; branchlets angular at first. New growth reddish.
Foliage	Blue-green, firm, elliptical phyllodes 2·5–6 cm × 8–20 mm, somewhat similar in shape to those of *A. myrtifolia* Willd.; with prominent mid-nerve closer to upper margin, branching minor veins, ending in a soft curved tip; margins lighter and slightly thickened; reducing at base into a small smooth stalk which appears to run down into the stem; a prominent gland 2–4 mm from base.
Flowers	Numerous bright lemon-yellow balls 5–6 mm diameter, each of 13–15 flowers, on smooth stalks 3–4 mm long in axillary racemes of up to 30 flowers, longer than the phyllodes; small triangular bracts at base. Flowering October.
Pods	Oblong, flat, pinkish-brown, 2·5–4·4 cm × 7–10 mm with lighter nerve-like margins, straight or slightly constricted between seeds. The seeds are often parasitised by insects.
Seeds	Grey-brown to black, oval, flattened 4 mm long, longitudinal in pod; seed-stalk thickening into a boat-shaped aril.
Identification	Colour and veining of phyllodes, position of gland, long racemes of flowers and flattened pods.
Comments	Shrub appears to sucker freely. It has adapted well to Tas. gardens in well-drained sites and is known to withstand snow and heavy frosts. It is grown from cuttings, suckers or seeds.

Acacia pataczekii

60 *Acacia podalyriifolia*

Group 8

A. Cunn. ex G. Don

Common name	Mt Morgan or Queensland Silver Wattle
Meaning of name	With leaves like a South African plant, *Podalyria*.
Distribution	Found on sandy, well-drained sites in coastal or near coastal areas of Qld between Brisbane and Rockhampton, on Blackdown Tableland and nearby ranges; a garden escape in many areas.
Habit	Silvery foliaged tall shrub or small tree 3–5 m tall, often spreading as wide, with smooth grey bark and hairy, glaucous branches. Young stems, phyllodes and buds velvety with usually very dense white hairs.
Foliage	Silvery blue-green, usually densely, softly hairy oval to elliptical phyllodes 2–5 cm × 10–20 (–27) mm, sometimes undulating, with penniveins and near central nerve; rounded at tip into a small curved blunt point; abruptly narrowed at base into a tiny hairy stalk; usually one or two tiny marginal glands, sometimes absent. Older phyllodes much less hairy.
Flowers	Numerous bright yellow balls 6–8 mm diameter, each of 20–30 flowers, on hairy stalks *c.* 5 mm long, in often branched axillary racemes of 10–20 flowers, 5–10 cm long. Flowering May–July.
Pods	Immature pods usually densely hairy; very flat, dark brown, usually softly hairy 4–9 (–12) cm × 10–20 mm, with raised margins, a little irregularly constricted between seeds.
Seeds	Black, oblong-oval, 6–7 mm × 3–4 mm, longitudinal in pod; rather long seed-stalk with short folds, the last one thickened under seed.
Identification	Glaucous, densely hairy phyllodes, position and number of glands, size of flowers and flat, wide pods.
Comments	Fast-growing species cultivated widely for many years especially for its foliage, which has become naturalised in many warmer regions. Tolerant of many soil types, including clay, as long as drainage is good. Flowers when quite small in some areas, but in colder climates may be sensitive to frost when young. Subject to sooty mould and borer attack, it is considered to be short-lived at 10–15 years. Light pruning after flowering is recommended to retain shape.

Acacia podalyriifolia

61 *Acacia jucunda*

Maiden & Blakely

Common name None known.

Meaning of name Lovely, pleasing.

Distribution Restricted to south-east Qld on dry ranges, open woodlands between near Eidsvold to Chinchilla–Tara area and west, on loams or clay loams.

Habit Upright, sturdy blue-green shrub or slender tree 2·5–8 m tall with grey, slightly fissured bark on older trees; branchlets angular, glaucous, usually very shortly hairy. Young tips velvety or hoary, usually covered with short hairs.

Foliage Blue-green, smooth or almost so, obliquely-oval to narrow lance-shaped phyllodes 4–6·5 cm × 9–20 mm with nerve-like margins, mid-nerve usually a little closer to top margin, fine penniveins; abruptly rounded at tip into a tiny point; gradually narrowed at base into a grey, usually hairy stalk 2–5 mm long; margins near base often shortly hairy; prominent gland nearby indenting margin.

Flowers Masses of bright yellow balls 6–7 mm diameter, each of 15–25 flowers, on short slender stalks 2·5–5 mm long, in smooth or shortly hairy axillary racemes of 10–30 flowers, 3–9 cm long. Flowering July–September.

Pods Dark brown, thin, flat, smooth pods covered with bloom 5–9 cm × 8–10 mm with almost straight, slightly thickened margins, irregularly narrowed between and slightly rounded alternately over seeds.

Seeds Black, oblong, 4–5 mm × 2–3 mm, longitudinal in pod; seed-stalk thickened at base into club-shaped aril.

Identification Somewhat resembles *A. podalyriifolia* A. Cunn., but differs mainly in habit, much less hairy, greener phyllodes, and position of basal gland.

Comments Attractive shrub for warmer climates but grows well in open, well-drained positions in southern states; seems to be slow to flower in Tas. but not affected by frost. It is recorded as being free-suckering.

Acacia jucunda

62 *Acacia uncifera*

Benth.

Common name None known.

Meaning of name Referring to hooked phyllodes.

Distribution Apparently restricted to several small sandstone areas of Great Dividing Range in Qld, mainly on Burra Range near Pentland; originally collected on the headwaters of the Nogoa River.

Habit Slender, much-branched hairy-leafed shrub to 5 m tall with rounded branchlets covered in dense white hairs.

Foliage Softly hairy, thick green, elliptical phyllodes 2·5–5 cm × 8–20 mm with central nerve and margins densely clothed with white hairs, ending in a fine firm point, curved upwards; narrowing at base into a short hairy stalk 1·5–2 mm long; small gland near base embedded in hairs, several others along top margin. Fine brown stipules at base of young phyllodes.

Flowers Numerous deep, bright yellow balls 8–9 mm diameter, each of 25–30 flowers, on densely hairy stalks 5–6 mm long, in racemes 5–8 cm long. Flowering June–August and later.

Pods Dull, felty, dark brown, thin, flat, 2–6·5 cm × 6–10 mm with slightly thickened margins, a little raised over, irregularly and slightly constricted between seeds. Scattered hairs on old pods.

Seeds Dull brown, marbled, round, oblique to transverse in pod; seed-stalk folded back and thickened into a small cap-like aril.

Identification Related to *A. podalyriifolia* and *A. jucunda*, but differs in shape of phyllodes, position and number of glands, and degree of hairiness.

Comments Spectacular shrub suited more to a sunny, well-drained position in drier northern regions. It is not considered consistently reliable in some soils in Atherton, and is reported to be difficult to grow in Rockhampton. Grown from seeds.

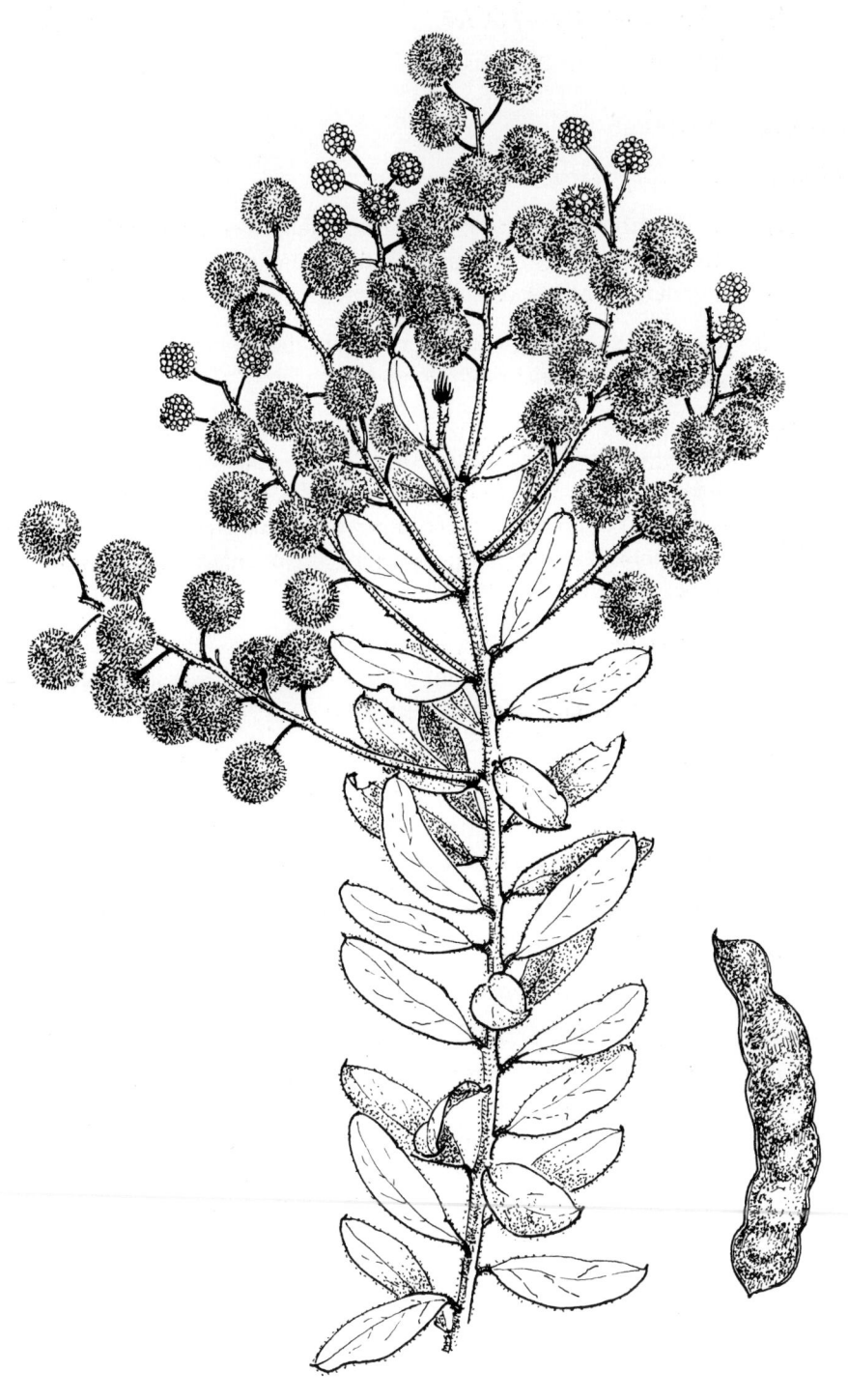

Acacia uncifera

63 *Acacia pyrifolia*

DC

Group 8

Common name	Ranji Bush
Meaning of name	Referring to pear-shaped phyllodes.
Distribution	WA northern districts of Fortescue, Ashburton and Carnegie; from north coast as far south as Meekatharra and east into the desert; common on hills and plains in sandy, gravelly soils.
Habit	Usually a straggly, glaucous, prickly shrub to 4 m × 3 m with smooth greyish bark; branches ± round, finely ribbed, often covered with bloom.
Foliage	Blue-green, thick, very stiff, large ± pear-shaped phyllodes 3·5−7·5 cm × 3−4·5 cm, with wavy, thickened margins, penniveined, prominent central nerve, ending abruptly in a long, sharp point 2−3 mm long; narrowing at base; gland present; stipules spiny, spreading.
Flowers	Numerous, dense, bright yellow balls 8−10 mm diameter, each of 70−80 flowers, on glaucous stalks 8−10 (−13) mm long, often two together, in long racemes of 10−12 flowers. Flowering April–October.
Pods	Stalked, light orange-brown, papery, with slight bloom, curved, flat 4−8 cm × 10−12 mm, rounded over seeds, margins thickened, irregularly constricted between seeds.
Seeds	Dull, dark brown, round-oval 5−7 × 4−5 mm transverse in pod; seed-stalk yellowish, folded twice, thickened into a small boat-shaped aril.
Identification	Large, pear-shaped phyllodes with long, sharp point; spiny stipules; flower-heads with numerous flowers in long racemes.
Comments	Not well known in cultivation; it is very slow-growing in Melbourne. There are two species which may be confused with *A. pyrifolia*: *A. inaequilatera*, a larger species of WA and NT, with smaller phyllodes, main nerve off-centre; buds and racemes purplish. *A. strongylophylla* from SA and NT, a small shrub with smaller, rounded phyllodes with central nerve, fewer flowers in smaller flower-heads.

Acacia pyrifolia

64 *Acacia iteaphylla*

F. Muell. ex Benth.

Common name	Flinders Range or Willow-leafed Wattle
Meaning of name	Referring to willow-like phyllodes.
Distribution	Confined to SA in mainly Flinders and Gawler Ranges in 250 mm annual rainfall area.
Habit	Several forms are recorded, one upright, the other pendulous, bushy, silvery blue-green shrubs 2–4 m tall and of the same width with smooth, long slender green branches; branchlets smooth, flattened; new growth is pinkish to purplish tipped.
Foliage	Silvery blue-green, slender, narrow lance-shaped phyllodes 5–14 cm × 3–8 mm with lighter margins, central nerve ending in a long fine straight or curved point; tapering at base into a smooth curved stalk; gland near base.
Flowers	Masses of pale or bright lemon-yellow perfumed balls 5–6 mm diameter, each of 8–17 flowers, on smooth short stalks in racemes of 8–12 flowers. Buds enclosed in conspicuous large, brown-tipped bracts similar to those of *A. suaveolens* and *A. subcaerulea*. Flowering April–September, sometimes earlier.
Pods	Conspicuous bunches of flattened blue-green pods drying brown, 6–12 cm × 8–10 mm with raised lighter margins, alternately rounded over and irregularly constricted between seeds.
Seeds	Dull black, oval 5–6 mm × 3 mm, longitudinal in pod; seed-stalk with short folds thickening into fleshy aril at side of seed.
Identification	Slender silvery phyllodes, conspicuous bracts enclosing buds, pale flowers and masses of blue-green pods.
Comments	Grown from cuttings and seeds. A hardy, adaptable, fast-growing decorative shrub, possibly one of the most widely grown acacias in southern Australia. It is frost tolerant, reportedly moderately salt tolerant but grows best in full sun in a well-drained position. A little pruning may be necessary after flowering, but hard pruning is not recommended.

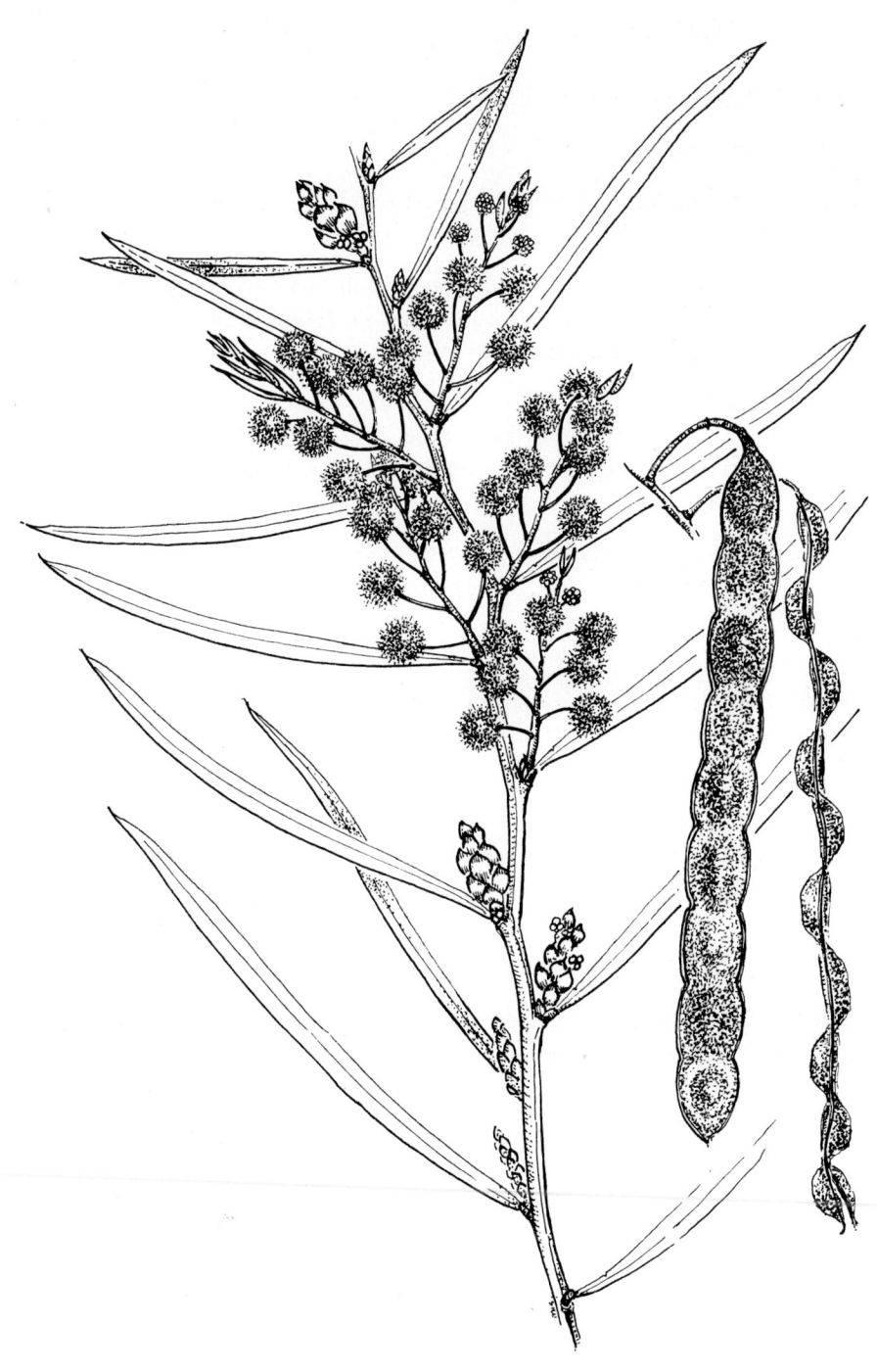

Acacia iteaphylla

65 *Acacia suaveolens*

(Sm.) Willd.

Group 9

Common name	Sweet-scented Wattle
Meaning of name	Sweet smelling.
Distribution	Common and widespread in sandy soils of east coast and tablelands of Qld, NSW, Vic., Tas. and south-east SA.
Habit	Slender, erect, smooth blue-green shrub 1–3 m tall often with few branches, sometimes straggly; branchlets smooth, often glaucous, acutely angled and flattened. New growth often pinkish.
Foliage	Blue-green, thick, smooth, slightly curved phyllodes 5–15 (–17) cm × 3–10 mm with nerve-like margins, central nerve ending in a long curved or straight point; tapering at base into a smooth wrinkled stalk; small gland at or near base, sometimes another at tip.
Flowers	Buds at first encased in conspicuous scaly deciduous bracts; fragrant, pale lemon-yellow balls 5–8 mm diameter, each of (3–) 6–10 flowers, on short, smooth stalks 2–5 mm long, in often crowded, short, axillary racemes. Flowering April–September.
Pods	Stalked, conspicuous, bluish, oblong flat, usually straight pods drying brown, 2–5 cm × 10–20 mm with thickened margins, little if at all constricted between seeds.
Seeds	Shiny black, oblong 6–7 mm × 2·5–4·5 mm, transverse in pod; short seed-stalk folded several times and thickened into a small oblique aril.
Identification	Buds encased in deciduous bracts, pale flower-heads each of few flowers, flat pods and very early flowering.
Comments	Hardy shrub suitable for coastal, near coastal and some inland gardens in light soils. It is considered to be short-lived in some areas. Grown from cuttings or seeds.

Acacia suaveolens

66 *Acacia gillii*

Group 10

(Maiden) Maiden & Blakely

Common name   Gill's Wattle

Meaning of name   Named for W. Gill (1851–1929), a conservator of forests who collected extensively in SA.

Distribution   Confined to loamy soils in open scrub on Eyre Peninsula between Flinders highway and coast, SA.

Habit   Slender, open, wiry shrub or small tree to 4 m tall with smooth grey bark and round brown pendulous branches; smooth, often zig-zagging, flattened angular branchlets.

Foliage   Green, smooth, leathery, spreading, often reflexed phyllodes 5–15 cm × 3–6 (–12) mm with lighter margins and fine laterals; prominent central nerve ending in a rounded tip; tapering at base into a smooth, curved, wrinkled stalk; gland near base.

Flowers   Dense bright yellow balls 7–8 mm diameter, each of 30–50 flowers, on smooth thickish stalks 10–20 mm long, singly or in simple zig-zagging racemes of up to 12 flowers. Flowering August–December and at other times.

Pods   Stalked, dull brown to fawn, usually curved 6–17 cm × 4–7 mm with thickened margins, rounded over and slightly constricted between seeds; pods in clusters.

Seeds   Black, 4 mm × 2 mm, longitudinal in pod; red-brown seed-stalk in double fold around seed, thickening into a conspicuous aril.

Identification   Previously known as a variety of *A. retinodes* Schldl. but it differs in its zig-zagging pendulous branchlets, more rigid phyllodes, different flowers and more woody pods.

Comments   Grown from seeds; it appears to be adaptable to many conditions and is growing successfully in Vic. and Tas. in warm, well-drained positions. It has been listed as an 'endangered' species in SA.

Acacia gillii

67 *Acacia rivalis*

Group 10

J. M. Black

Common name	Silver Wattle
Meaning of name	Growing near water.
Distribution	Confined to northern Flinders Range, SA, and to south-west NSW, on ridges, rocky hillsides and near water-courses.
Habit	Many-branched, umbrella-shaped shrub or small tree 3–5 m tall with reddish bark slightly fissured at base and with often drooping branches; branchlets smooth, round, angular at tips, sometimes covered with bloom. New growth reddish clothed with scattered, moderately dense white hairs.
Foliage	Green, smooth, shining, curved, narrow lance-shaped phyllodes 4–14 cm × 2–5 mm with a single central nerve, narrowing into a small, blunt, usually recurved tip; tapering at base into a small flattened stalk; gland some distance along top margin.
Flowers	Small bright yellow balls 5–6 mm diameter, each of 30–40 flowers, on hoary single jointed stalks 5–10 mm long or in short fine hoary racemes of 4–10 flowers. Buds golden hairy. Flowering May–November.
Pods	Dark brown, smooth, narrow, beadlike, straight or curved, 7–13 cm × 3–5 mm, rounded over and constricted between seeds.
Seeds	Shining, oval-oblong, longitudinal in pod; fairly long seed-stalk folded once before half encircling seed ending in a thickened aril.
Identification	Shining smooth phyllodes, placement of gland, flower characteristics; somewhat similar to *A. calamifolia*.
Comments	Grown from seeds; a plant suited to hot, dry inland gardens. It is often found growing in association with *A. victoriae*, *A. tetragonophylla*, *Hakea edniana* and several *Eremophila* species.

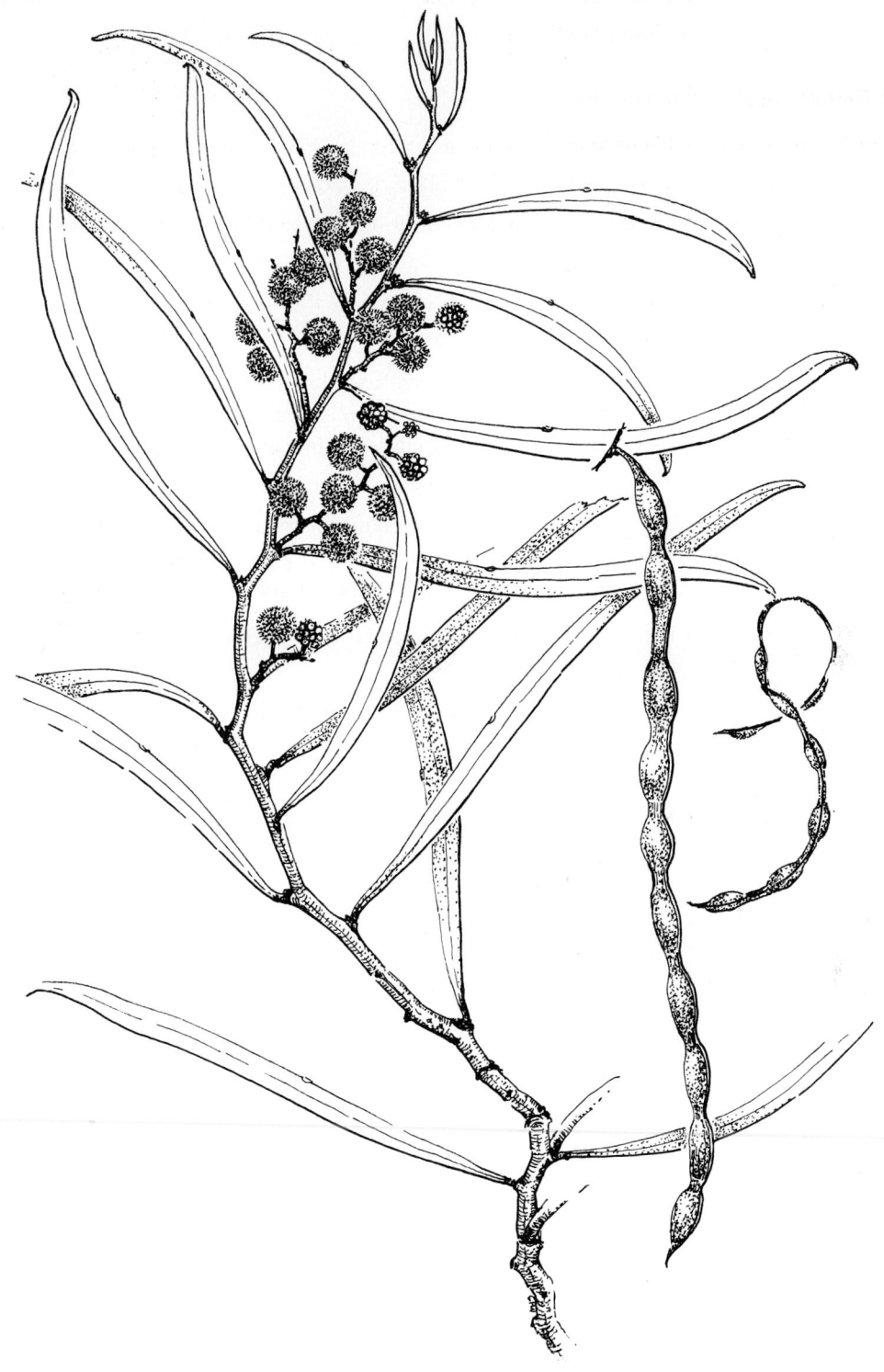

Acacia rivalis

68 *Acacia chrysella*

Maiden & Blakely

Common name None known.

Meaning of name Golden yellow; presumably referring to minute golden pubescence of flower stalks.

Distribution WA south-west in Bencubbin, Leonora and Esperance areas; usually in sand or sandy loam.

Habit Variable, dense bushy shrub to 3 m × 2–4 m; branchlets without hairs.

Foliage Green, sometimes blue-green, usually incurved, flat, linear phyllodes 4–13 cm × 1–4 (–5·5) mm with central nerve rarely prominent; tapering at tip into a long curved point; gland 1–3 cm from base, occasionaly two glands present.

Flowers Light golden balls 3–4 mm diameter, each of 15–25 flowers, on stalks 2–4 mm long, in slender, sparsely hairy racemes of 3–10 flower-heads, mostly 4–20 mm long. Flowering December–August.

Pods Blackish to 10 cm × 5–6 mm rounded over and irregularly constricted between seeds.

Seeds Oblong 4–6 mm × 2–3 mm longitudinal in pod; seed-stalk light to red brown, half to entirely encircling the seed in a single fold, ending in a thick aril.

Identification Close to *A. aestivalis* and *A. harveyi* but differs in its narrow, usually incurved phyllodes, gland position, light golden flowers and narrowly constricted pods.

Comments This is a complex species which is being revised for future publication.

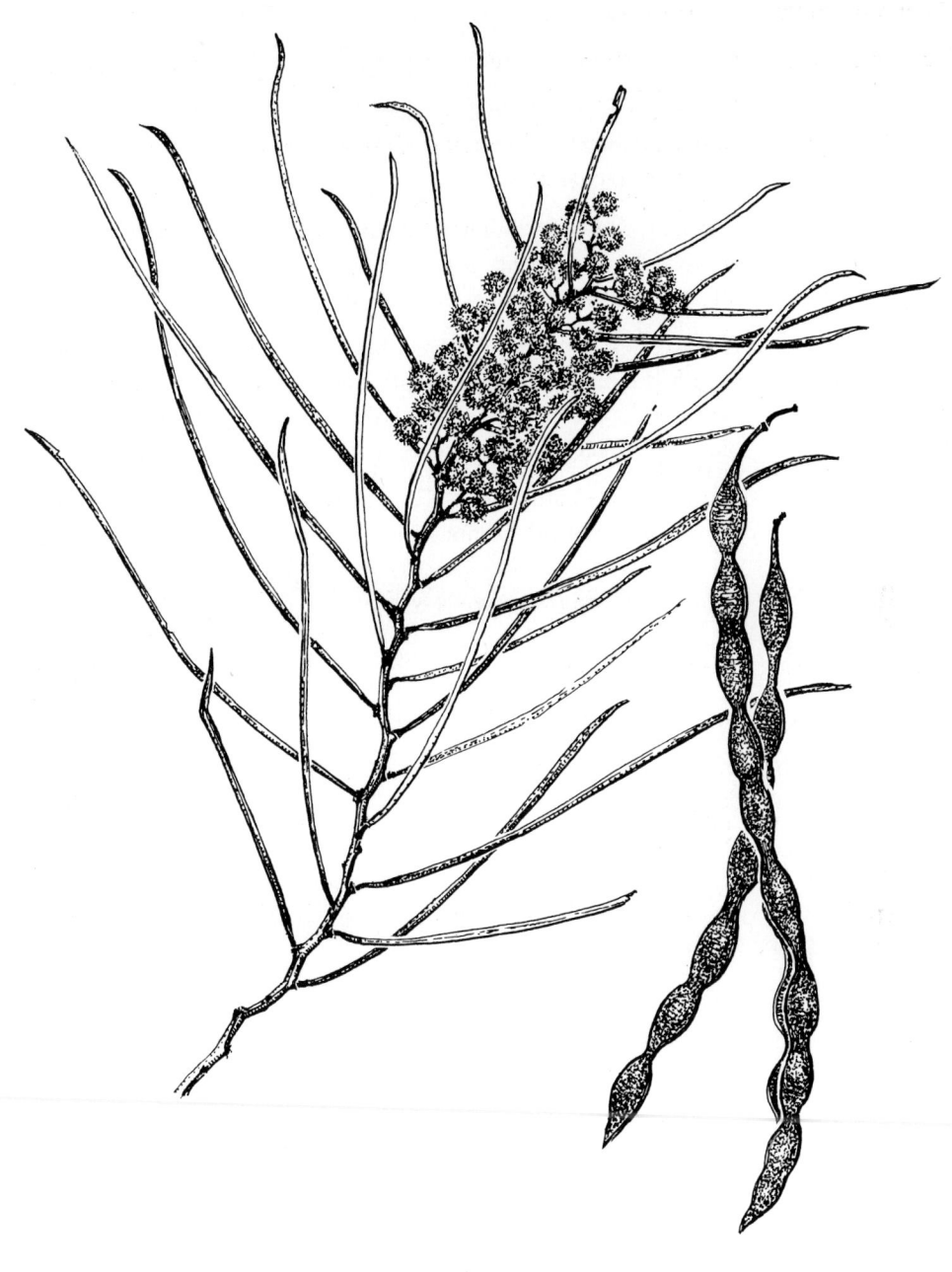

Acacia chrysella

69 *Acacia rostellifera*
Benth.

Common name	None known.
Meaning of name	Having little beaks; presumably referring to the phyllode tips.
Distribution	WA in Irwin and Darling districts, from Shark Bay to Esperance (not common along south coast); common on Rottnest Island; mostly restricted to coastal dunes where it may form thickets.
Habit	An erect, widely-spreading shrub or small tree 2–5 m × 3–5 m, often with a willow-like habit; branchlets angular, brownish, yellow ribbed, often glaucous between ribs, zig-zagging.
Foliage	Green, thick, straight or curved, linear-lance-shaped phyllodes (4–) 5–12·5 cm × 4–9 mm with thickened, yellowish margins, minor veins obscure, a central nerve, rarely two, ending in a dark, recurved, blunt tip; small gland near base.
Flowers	Dense, bright gold balls 7–12 mm diameter, each of 25–30 flowers, on ± stout stalks 4–9 mm long, in short racemes of 2–8 flowers. Plants flower rather infrequently, presumably due to their suckering habit. Flowering August–November.
Pods	Dark brown, rough surfaced, brittle, usually curved, flat 4–9 cm × 6–8 mm, with nerve-like margins, slightly raised over and evenly constricted between seeds.
Seeds	Black, oval-oblong 5–6 mm × 3·5–4 mm, longitudinal in pod with orange-red aril.
Identification	Green phyllodes, short racemes of very bright flowers; brittle, evenly constricted pods; occurring only in coastal dunes. Allied to *A. xanthina* which has blue-green, 2-nerved, often broader phyllodes and which is restricted to coastal limestone.
Comments	Hardy species suited to coastal and inland areas. It is being grown successfully in eastern states. Grown from cuttings and seeds.

Acacia rostellifera

70 *Acacia amoena*

Wendl.

Common name	Boomerang Wattle
Meaning of name	Pleasing, lovely.
Distribution	Coast, Dividing Range and central western slopes of NSW, and in Vic. confined to a small area of the Eastern highlands and banks of nearby rivers.
Habit	Mostly green-leafed shrub 2–3 m tall with mottled grey bark and smooth branches; young branchlets angular, soon becoming round.
Foliage	Smooth, dark green, obliquely lance-shaped, curved or straight phyllodes 2·5–7 cm × 6–12 mm with nerve-like margins; central nerve ending in a small straight or recurved blunt tip; tapering at base into a smooth stalk; 2–4 often prominent well-spaced glands on upper margin.
Flowers	Masses of small, dense bright yellow balls 4–5 mm diameter, each of 6–12 flowers, on short smooth stalks in racemes 3–5 cm long. Flowering August–October.
Pods	Dark brown, flat, straight or curved, 5–10 cm × 6–8 mm with nerve-like margins, rounded over but little constricted between seeds.
Seeds	Black, oblong-oval, 4–5 mm × 2–3 mm, longitudinal in pod; seed-stalk very long, completely encircling seed three times, then thickened into a fleshy aril.
Identification	Position and number of glands; seed-stalk encircling seed three times.
Comments	A hardy, frost tolerant shrub in southern states; will grow in a rocky exposed situation, in full sun or semi-shade in well-drained soils. Grown from seeds.

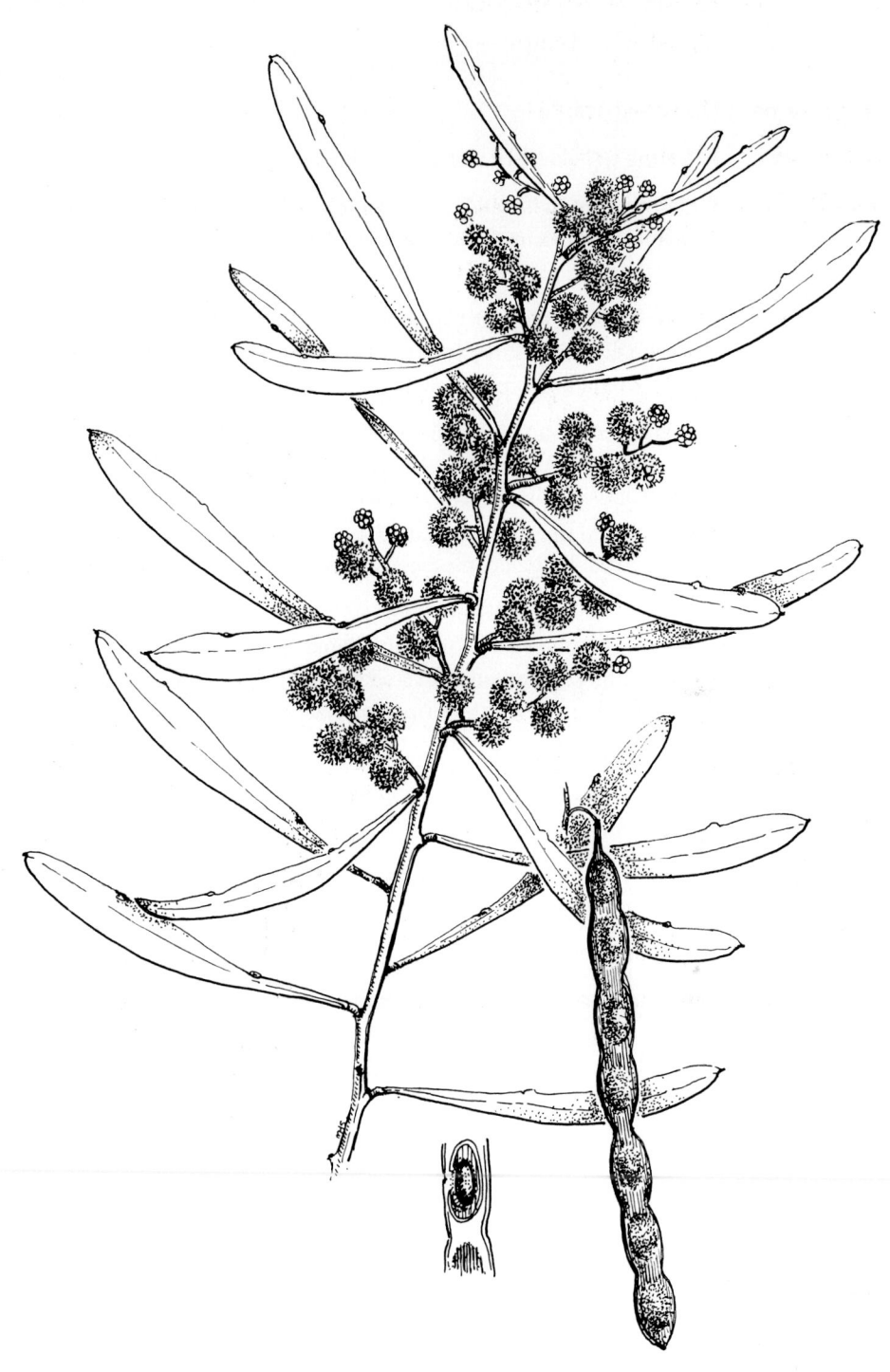

Acacia amoena

71 *Acacia hakeoides*

A. Cunn. ex Benth.

Group 10

Common name	Hakea-leaf Wattle
Meaning of name	Referring to hakea-like shape of phyllodes.
Distribution	Widespread on light sandy, occasionally heavier soils in the drier parts of mainland states, especially western slopes and plains of NSW, SA, not common in southern Qld, in north-western Vic. and eastern WA.
Habit	Usually a dense, spreading, smooth green-leafed shrub 1–4 m × 2–5 m (occasionally a small tree 5–10 m tall) with smooth, grey bark and smooth round branches; branchlets at first angular, often flattened.
Foliage	Variable, green, smooth rather thick, ± oblong, slightly curved phyllodes 4–12 (–14) cm × 5–12 (–15) mm, widest at top, with prominent central nerve and minor net veins, ending abruptly in a blunt rounded tip; narrowing gradually at base into a yellow-green stalk 1–2 mm long; a conspicuous gland about 5 mm from base to near middle of top margin.
Flowers	Dense bright yellow balls 6–8 mm diameter, each of 20–25 flowers, on rather thick, short, smooth stalks in axillary racemes of 6–12 flowers, 2–5 cm long. Flowering July–September.
Pods	Dark brown, smooth, beadlike, flattish, curved 4–10 cm × 4–7 mm with thickened lighter margins, much constricted and lengthened between seeds.
Seeds	Black, oval, 6–7 mm × 4–4·5 mm, longitudinal in pod; short seed-stalk thickened and widened into a fleshy aril.
Identification	Shape and venation of phyllodes, position of gland and beadlike pods.
Comments	Grown from cuttings and seeds. It often forms dense thickets and is reported to sucker freely. Used for shelter belts, it is drought resistant and moderately frost tolerant. It is grown in gardens and roadside plantations in southern states.

156

Acacia hakeoides

72 *Acacia rubida*

A. Cunn.

Common name	Red-stemmed Wattle
Meaning of name	Becoming red, referring to stems and leaves, especially when dry.
Distribution	Widely distributed in most eastern states excluding Tas., mainly on slopes, river banks, hills and mountains in 650 mm annual rainfall area; confined to granite belt near Stanthorpe in southern Qld.
Habit	Variable green, blue-green or reddish leafed bushy shrub or small tree 1·5–12 m tall with grey-brown bark, often fissured at base; branchlets smooth or with scattered hairs, angular and flattened, often reddish. Feathery bipinnate leaves often persistent on lower branches and present at times on flowering stems. New growth often red.
Foliage	Thick, smooth, green to blue-green sickle-shaped, straight or curved phyllodes 5–18·5 cm × (8–) 10–25 mm with often reddish thickened margins, prominent mid-nerve sometimes nearer top margin, penniveins; tapering to both ends, at tip into a long, often hooked, dull point and at base into a stout stalk 1–3 mm long; prominent gland some distance from base. Phyllodes and stems sometimes reddish; this becomes more pronounced on drying.
Flowers	Numerous deep bright yellow balls 5–6 mm diameter, each of 10–15 flowers, on usually smooth red stalks, in loose racemes of (8–) 10–15 flowers. Flowering July–October.
Pods	Red-brown, firm-textured, straight or slightly curved, (6·5–) 10–12·5 cm × 5–7 (–9) mm, margins thickened, flattened, little if at all constricted between seeds.
Seeds	Black, oval 5–6 mm × 2 mm, longitudinal in pod; seed-stalk long, encircling seed in a double fold.
Identification	Reddish stems and leaves becoming red when dry; retention of juvenile foliage.
Comments	Frost resistant and moderately drought tolerant species; grows successfully in a wide range of conditions. Grown from cuttings and seeds.

Acacia rubida

73 *Acacia gladiiformis*

A. Cunn. ex Benth.

Group 10

Common name	Sword-leaf Wattle
Meaning of name	Referring to sword-like shape of phyllodes.
Distribution	Restricted to ridges and sandy soils in Pilliga Scrub, Warrumbungles to Blue Mountains in NSW.
Habit	Slender, erect, smooth, green-leafed shrub with several stems 1–3 m tall, occasionally more, with smooth branches; branchlets angular, soon becoming round.
Foliage	Green, smooth, upward curved phyllodes 7·5–15 cm × (6–) 8–12 mm with single central nerve and thickened margins ending in a blunt, small, hooked point; narrowing at base into a smooth thick stalk; 2–4 prominent glands spaced along top margin.
Flowers	Dense large bright yellow balls 9–10 mm diameter, each of 40–50 flowers, on short, stoutish stalks 3–4 mm long, in often zig-zagging racemes shorter than phyllodes. Flowering June–September.
Pods	Black-brown, smooth, firm-textured, 5–11 cm × 4–6 mm with slightly thickened lighter coloured margins, rounded over and slightly narrowed between seeds.
Seeds	Black, oblong, 5 mm × 3 mm, longitudinal in pod; seed-stalk long encircling seed in a double fold and thickening into a small cap-shaped aril.
Identification	Sometimes confused with *A. hakeoides* but differs in habit and in longer more incurved phyllodes, with more glands and seed-stalk encircling seed.
Comments	Hardy species requiring a sunny, well-drained position in the south; will tolerate semi-shade in coastal SA. Grown from seeds.

Acacia gladiiformis

74 *Acacia beckleri*

Tindale

Common name	Barrier Range Wattle
Meaning of name	Named for Dr H. Beckler who made extensive collections in 1860s, and who has many species named for him.
Distribution	Scattered through semi-arid to arid areas of SA in the vicinity of Gawler and Flinders Ranges, and in western NSW on the bare hills of Barrier Range near Broken Hill to near Ivanhoe.
Habit	Erect or spreading bushy green or blue-green shrub 1–3 m × 2–3 m with almost round, slightly ribbed smooth branches, occasionally covered with bloom. New growth often red and shining.
Foliage	Thick leathery green to blue-green straight or curved lance-shaped phyllodes 7–17·5 cm × 6–25 mm with thickened margins, many fine penniveins and a prominent central nerve ending in a blunt or acute tip; tapering at base into a thick curved stalk; prominent glands vary from one to four along top margin.
Flowers	Large, usually dense, very bright yellow balls 10–17 mm diameter, each of 50–90 flowers, on short stout to extremely wide (to 3 mm) swollen, hairy, ribbed stalks, singly or in sturdy racemes of 3–7 flowers. Buds hairy. Flowering May–August, sometimes later.
Pods	Stalked, dark red-brown, smooth, 7–13 cm × 5–8 mm with lighter thickened margins, rounded over seeds and slightly constricted between them.
Seeds	Black, oblong 5–6 mm × 2–2·5 mm, longitudinal in pod; seed-stalk once folded thickening into a cup-shaped aril.
Identification	Resembles *A. notabilis* F. Muell. but differs in its longer narrower phyllodes, usually several glands, larger flowers, often extremely thick flower-stalks and longitudinal seeds.
Comments	Widely grown, hardy, adaptable to a wide range of conditions in an open, warm, well-drained position; subject to tip damage from frost in northern Tas. Light pruning after flowering recommended.

Acacia beckleri

75 *Acacia pycnantha*
Benth.

Common name	Golden Wattle
Meaning of name	Referring to dense flowering of shrub.
Distribution	Widespread, sometimes common, on dry stony ground or sand, from as far west as southern Eyre Peninsula in SA across Vic. and into southern regions of NSW. Naturalised in many places on tablelands NSW and on east coast of Tas.
Habit	Sturdy, smooth, shining green, pendulous-leafed bushy shrub or small tree with spreading crown 4–8 m tall, with dark brown or grey rather smooth bark, often rough at base of trunk; branchlets almost round, reddish, sometimes covered with bloom. New growth often a bronze colour.
Foliage	Variable, smooth, thick, green, curved, broadly lance-shaped pendulous phyllodes 6–20 cm × (5) 8–50 mm usually widest above centre, with prominent margins, lateral veins and mid-nerve, ending in an elongated, blunt tip; tapering gradually at base into a thickened reddish stalk *c.* 7 mm long; prominent raised gland near or a distance from base; occasionally a second gland present. Phyllodes often largest when young.
Flowers	Masses of fragrant, large, dense bright yellow balls 8–10 mm diameter, each of 50–80 flowers, on smooth stout stalks 5–6 mm long, in short, stout axillary racemes of 6–12 flowers or in panicles. Flowering August–October.
Pods	Dark brown, thin-textured, flattened, straight or slightly curved, 5–12 cm × 5–7 mm with thickened margins, rounded over and slightly constricted between seeds. Pods bright green when young.
Seeds	Black, oblong 4–5 mm × 2·5–3 mm, longitudinal in pod; short seed-stalk occasionally folded and thickened upwards into a small fleshy aril.
Identification	Position and shape of gland(s) on phyllode; dense large flower-heads in racemes; longitudinally placed seeds. A pale-flowered variety is widely grown, which flowers from April–August.
Comments	Useful as a shade, shelter or garden specimen. It is drought resistant and grows in a wide range of soils including coastal areas with some protection from salt laden winds. Young plants are rather frost tender. It grows very quickly and is often considered to be short-lived (8–10 years) in cultivation. Grown from cuttings and seeds. Previously considered a valuable source of tannin bark and has been grown overseas for this purpose.

Acacia pycnantha

76 *Acacia notabilis*
F. Muell.

Common name	Notable Wattle
Meaning of name	Noteworthy.
Distribution	Occasional or common in open woodlands over a wide area of SA and the western plains of NSW around Broken Hill, in 250 mm annual rainfall areas.
Habit	Blue-green bushy shrub usually branching from ground level 1–3 m (sometimes taller) × 4–6 m wide with smooth grey-brown bark and round branches. Young stems angular, flattened, sometimes reddish.
Foliage	Leathery blue-green straight or curved oblong-lance-shaped phyllodes 5–15 cm × 5–25 mm with thickened margins, fine laterals and a prominent central nerve often nearer top margin; rounded or elongated at tip, blunt; narrowed at base into a curved wrinkled stalk; small gland near base.
Flowers	Large, dense, mid to bright yellow balls 8–9 mm diameter, each of 50 or more flowers, on thick smooth stalks 6–7 mm long in short racemes of 4–15 flowers, mostly shorter than phyllodes. Flowering July–November.
Pods	Dark red-brown, straight, flat, oblong, often covered with bloom, 3–7·5 cm × 8–12 mm with thickened nerve-like margins, closely raised alternately over seeds, occasionally and irregularly constricted between them.
Seeds	Dull black, oval, 5–7 mm × 4–5 mm, transverse in pod; long threadlike seed-stalk encircling seed in a double fold, ending in a small fleshy aril.
Identification	Allied to *A. validinervia* Maiden & Blakely, but differs in venation of phyllodes, broader pods and transverse seeds.
Comments	Hardy shrub for hot dry inland areas, but it is growing successfully in SA, Vic. and Tas. gardens; thought to be reasonably salt, clay and lime tolerant; a recommended plant for Alice Springs area; grown from seeds.

Acacia notabilis

77 *Acacia microbotrya*

Benth.

Group 10

Common name	Manna Wattle
Meaning of name	Small bunch of grapes; referring to the flower-heads arranged in small racemes.
Distribution	WA south-west, in Irwin, Avon, Darling, Stirling, Eyre and Coolgardie districts, from north of Geraldton south to Bremer Bay and east to Kalgoorlie; widespread.
Habit	Bushy shrub or small tree 2–7 m × up to 5 m wide with grey bark, rough at base; branchlets slightly angular at first.
Foliage	Often grey-green, pendulous, curved, lance-shaped phyllodes, extremely variable in shape and size, mostly 7–13 cm × 6–15 mm with nerve-like margins, minor penninerves, prominent mid-nerve, tapering to a short, curved point; tapering at base into an orangey stalk; gland small, occasionally two, on top margin, sometimes absent.
Flowers	Numerous sprays of small, bright yellow balls 5–6 mm diameter, each of about 20–30 flowers, on short, golden-hairy stalks 3–4 mm long, in short racemes of 3–4 or more flowers, 1·3–4 cm long. Flowering March–June, also irregularly throughout the year.
Pods	In clusters, dark brown, finely veined, bead-like, 10–20 cm × 7–8 mm with slightly thickened margins, alternately raised over and much constricted between seeds.
Seeds	Black, oblong, flat, 5–8 mm × 4–5 mm, as many as 10 seeds longitudinal in pod; seed-stalk long, slender, folded 2 or 3 times and thickened below seed into a large, club-shaped, fleshy aril.
Identification	Numerous small flower-heads in racemes, veining of phyllodes, bead-like, long pods.
Comments	Reported to be hardy, drought and frost tolerant, fast growing in warmer areas; it is growing quite well in Tas. Useful for low shelter and as a source of honey or pollen; often flowers when quite small. A new *Acacia* has been separated from *A. microbotrya*; the name and details will be published soon.

Acacia microbotrya

78 *Acacia sclerosperma*

F. Muell.

Common name None known.

Meaning of name Referring to the hard seeds.

Distribution WA north-west in Irwin, north to Fortescue district, between Geraldton and Port Hedland, near salt lakes, muddy flood plains and red soils in low scrub.

Habit Shrub or small tree 1·3–6 m × 3–4 m, often spreading widely, with rough, grey bark at base; branchlets often reddish-brown or purplish-grey, somewhat angular at first. Young growth is often bluish-green.

Foliage Light green or blue-green, rather thick, spreading, curved phyllodes 5–12·5 cm × 1·5–4 mm with prominent mid-nerve, ending in a blunt, straight or hooked point; small gland some distance from base.

Flowers Strongly perfumed, large, loose, bright yellow balls 10–13 mm diameter, each of about 20 flowers, on conspicuous stalks 7–15 (–20) mm long, solitary or in very short racemes of 3–4 flowers, sometimes growing out into a leafy shoot. Flowering June–September.

Pods Large, dark brown, woody, often rough surfaced, bead-like pods 10–12 cm × 10–15 mm with thickened margins, much rounded over and constricted between seeds. Immature pods often with bloom.

Seeds Large, yellow-brown-black, shining, oval-round, 8–10 × 5–8 mm × 4 mm thick, longitudinal in pod; seed-stalk brown, short, ending in a small, red cap.

Identification Large pale seeds; woody, bead-like pods and narrow phyllodes; large flower-heads.

Comments Used for windbreaks and suitable for planting in inland and coastal areas, where it is reported to tolerate salt spray.

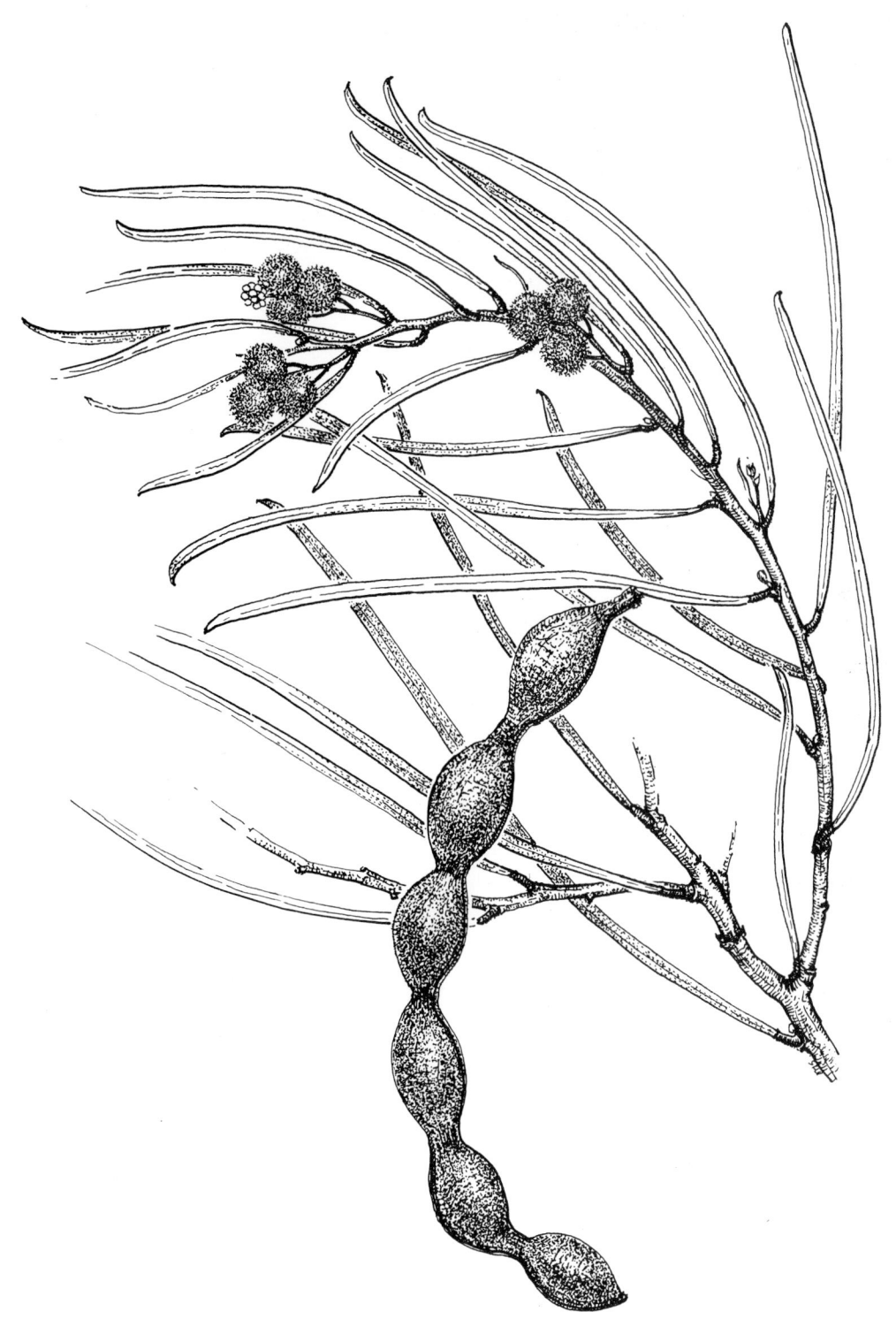

Acacia sclerosperma

79 *Acacia neriifolia*

A. Cunn. ex Benth.

Group 10

Common name	Silver Wattle, sometimes Bastard Yarran
Meaning of name	With Oleander-like leaves.
Distribution	Common, often on granitic soils of New England Tablelands of NSW, into southern Qld, near Inglewood and forests near Chinchilla, north through Isla Gorge and Blackdown Tableland.
Habit	Variable blue-green or dull green-leafed, slender shrub or small tree to 8 m tall with smooth, occasionally rough, grey bark; branchlets slender, smooth, angular, covered with white appressed hairs or bloom when young.
Foliage	Blue-green or green, straight or slightly curved, linear-lance-shaped phyllodes (5–) 7·5–17 (–21) cm × (3–) 5–10 (–17) mm, with appressed white hairs especially when young, with central nerve, fainter penniveins; margins prominent; tapering into a small blunt tip, often recurved; narrowing at base into a thickened, smooth, light green stalk 1·5–4 mm long; 1–3 glands, one near base, others spaced along top margin, occasionally one at tip.
Flowers	Numerous very bright yellow balls 7–8 mm diameter, each of 20–40 flowers, on slender, usually golden or white hairy stalks 3–5 mm long, in racemes of 13 or more flower-heads, 4–9 cm long. Flowering June–September.
Pods	Bunches of stalked, light red-brown, smooth, flat, slightly curved pods 8–26 cm × 6–10 mm with lighter coloured, thickened margins, raised alternately over seeds and very slightly narrowed between them.
Seeds	Large black rectangular, 5–8 (–12) mm × (4–) 6–9 mm, longitudinal in pod; seed-stalk fine, folded once before thickening into a large boat-shaped aril.
Identification	Hairy young phyllodes, multiple glands, large pods and seeds.
Comments	Frost resistant, moderately drought tolerant species. Sometimes used in windbreaks; grown successfully in eastern states including Tas. Grown from seeds. Timber is close-grained, tough, dark with light yellow heart-wood. Sometimes used as emergency stock fodder.

Acacia neriifolia

80 *Acacia penninervis*
Sieber ex DC.

Common name	Hickory Wattle or Mountain Hickory
Meaning of name	Referring to minor nerves which run from mid-nerve to margin of phyllodes.
Distribution	Widespread and common on coast, slopes and tablelands and somewhat inland in Qld, in NSW, ACT and in Vic. confined to an area near Seymour.
Habit	Variable, smooth-leafed shrub or small tree 3–12 m tall with grey, often mottled bark; branchlets smooth, angular, sometimes reddish. New growth often bright or dull red.
Foliage	Variable, green or blue-green smooth, straight or curved, oblong to broadly lance-shaped phyllodes 5–12 cm × (6–) 8–25 (–50) mm, usually widest above centre, with fine penniveins, nerve-like margins and prominent mid-nerve ending abruptly in a blunt rounded or longer tip; tapering at base into a smooth, curved stalk 2–5 mm long. Often a short secondary nerve terminating in a prominent gland on upper margin, some distance from base; a second gland occasionally present.
Flowers	Cream or pale yellow balls 7–8 mm diameter, each of 20–30 flowers, on smooth or hairy stalks 4–9 mm long, in loose terminal racemes or panicles. Flowering November–February, sometimes later.
Pods	Dull brown, thick, flat, straight or curved, 5–13 (–20) cm × 10–15 (–17) mm with thickened margins, rounded over and irregularly constricted between seeds.
Seeds	Black, oval, 6–7·5 mm × 3·5–4 mm, longitudinal in pod; seed-stalk short, folded once before widening into a large aril.
Identification	Gland on top margin of phyllode connecting by a secondary nerve to mid-nerve; pale flowers in racemes; large flat wide pods.
Comments	Widely grown, hardy small tree, suitable for windbreaks and farm forests; frost and snow tolerant; grown from seeds. The bark is considered suitable for pulping in paper making. It is recorded as being one of the finest acacia hardwoods, almost identical in texture and colour with the much-prized *A. melanoxylon* (Blackwood).

Acacia penninervis

81 *Acacia macradenia*
Benth.

Common name	Zig-zag Wattle
Meaning of name	Referring to elongated gland on phyllode margin.
Distribution	Common on sandy or stony soils on rocky hills, near creeks and along roadsides mainly in Leichhardt, Maranoa and Darling Downs districts, e.g. Blackdown Tableland, Chinchilla to Alpha–Tambo areas, Qld.
Habit	Spreading, smooth, green-leafed shrub or small slender tree 3–5 m × 3–4 m with smooth grey or brown bark and often prominently zig-zagging, pendulous, angular branchlets. New growth often red.
Foliage	Spreading or reflexed thick, smooth, curved green phyllodes (8–) 13–25 cm × (8–) 10–25 mm with prominent margins, penniveins and central nerve tapering into a long slender point; narrowing at base into a smooth, wrinkled, curved stalk, 2–3 mm long; a large elongated gland near base, occasionally one or two smaller glands along top margin; persistent, hard, reflexed stipules up to 5 mm long.
Flowers	Large, bright yellow, fragrant balls 9–10 mm diameter, each of 35–50 flowers, on usually smooth stalks, 3–5 mm long, in long, often twin, racemes of 8–15 flowers. Flowering July–August, sometimes later.
Pods	Clusters of stalked, dark brown, smooth, beadlike pods 7–9 cm × 4–5 mm raised over seeds alternately, and much narrowed between seeds. Immature pods reddish.
Seeds	Oblong-oval, black, 4·5–5 mm × 2–2·5 mm, longitudinal in pod; seed-stalk folded back and ending in a bright-coloured, boat-shaped aril on side of seed. Seeds collected October.
Identification	Arrangement and length of phyllodes, sharp persistent stipules, often zig-zagging branchlets.
Comments	A decorative, often cultivated plant well suited to warmer subcoastal or coastal areas; it requires good drainage and usually full sun; tolerates moderate but not severe frosts. Successfully grown in gardens from Townsville to Melbourne, but winters in Canberra have proved too severe.

Acacia macradenia

82 *Acacia bancroftii*

Maiden

Group 10

Common name	None known.
Meaning of name	Named for Dr T. L. Bancroft (1860–1933) medical officer at Eidsvold, a collector of native plants.
Distribution	Confined mainly to stony hillsides and ridges of South Kennedy, Leichhardt and Burnett districts from Clermont–Kingaroy, west to Alpha–Tambo districts, Qld.
Habit	Slender glaucous tree 5–6 m tall with an open crown, smooth pinkish-grey or brown bark; branchlets smooth or glaucous ± rounded. Juvenile phyllodes are often very large and new growth reddish.
Foliage	Smooth, usually blue-green, flat, sickle-shaped phyllodes 9–22 cm × 2–8·5 (–12) cm, often with one gland near base or halfway along, or with several conspicuously raised stalked glands along top margin; prominent mid-nerve nearer top margin and nerve-like penniveins; tapering at tip into a soft blunt point, and at base into curved, large, thick stalk 5–15 mm long.
Flowers	Large, dense bright yellow balls 7–10 mm diameter, each of 20–40 flowers, on smooth or occasionally hairy stalks 3–5 (–8) mm long, in racemes of up to 25 or more flowers, 7–16 cm long. Flowering May–July.
Pods	Clusters of stalked, flat, smooth, straight or curved pods, usually covered with bloom and transversely veined, 9–20 cm × 10–17 mm with thickened margins, raised over and occasionally irregularly constricted between seeds.
Seeds	Black, oblong, 6–10 mm × 4–5 mm × 2 mm thick, longitudinal in pod; wide pale seed-stalk completely encircling seed, ending in a club-shaped aril.
Identification	Usually blue-green phyllodes, number of often-raised, conspicuous glands, position of mid-nerve and long racemes of bright yellow flowers. Only a small proportion of phyllodes have stalked glands.
Comments	Suited best to drier northern areas; it is growing successfully in coastal gardens from Brisbane to Townsville, and in some areas of NSW. It is grown from seed.

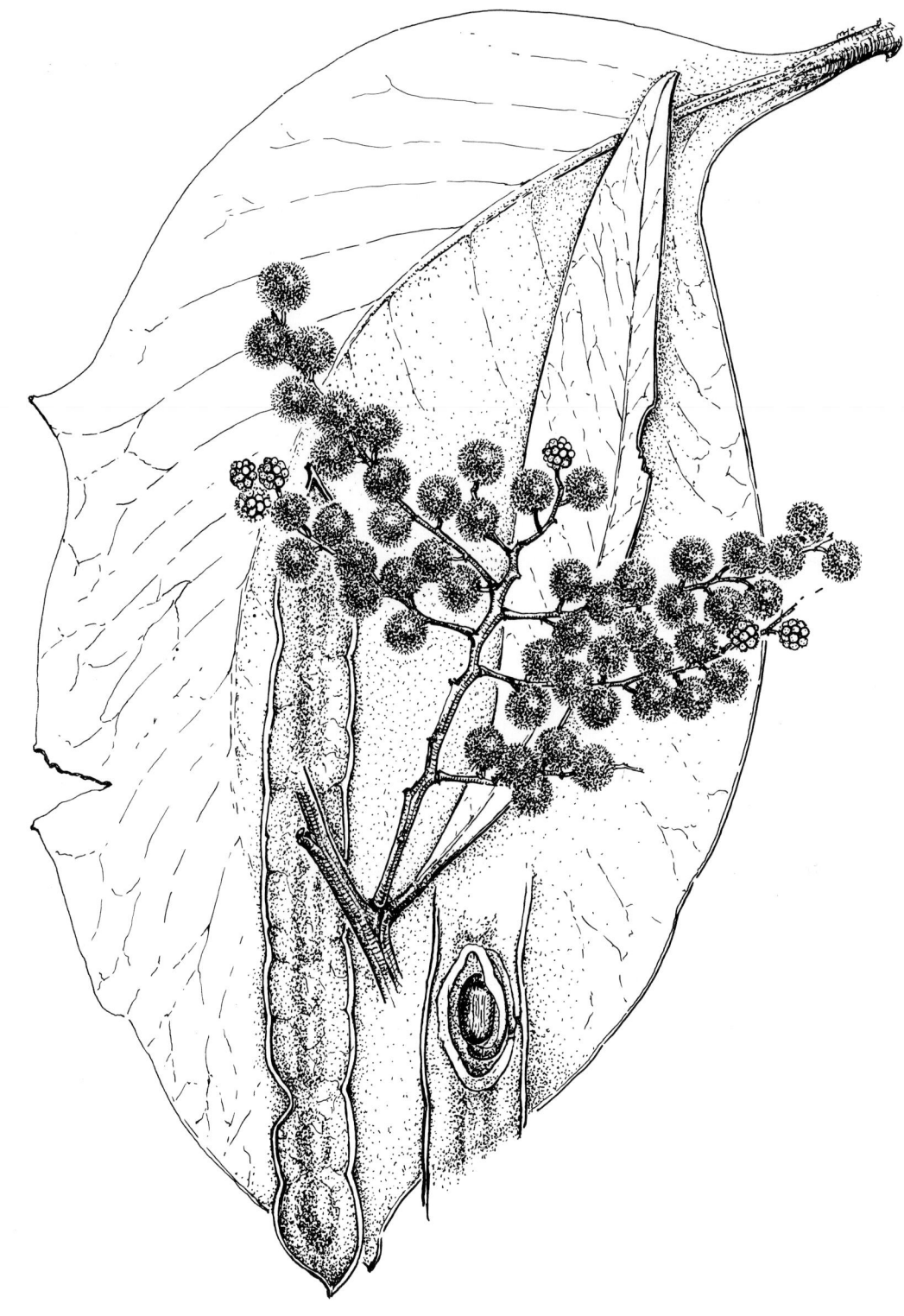

Acacia bancroftii

83 *Acacia victoriae*
Benth.

Group 10

Common name	Elegant or Bramble Wattle; Gundabluey in Queensland and New South Wales
Meaning of name	Named for Victoria River (Barcoo) in Queensland where the plant was first found.
Distribution	Widespread in open wood and shrublands, sometimes forming dense thickets on plains, on ridges and depressions and on creek banks in drier parts of all mainland states; rare in north-west Victoria.
Habit	Variable, usually spiny, light green, bushy shrub or small tree to 5 m tall and widespreading, much branched from near ground level with usually grey bark, fissured at base; branchlets smooth or hairy slightly angular, sometimes ribbed.
Foliage	Variable green or blue-green, linear to oblong, lance-shaped smooth or hairy phyllodes 2–6 (–8) cm × (1·5–) 2–8 mm with prominent central nerve ending abruptly at tip in a blunt usually hooked point; tapering at base into a tiny stalk; gland near base; two very sharp spiny stipules to 10 mm long, at base, sometimes absent.
Flowers	Perfumed pale to mid yellow balls 5–7 mm diameter, each of 15–30 flowers, on often hairy slender stalks 6–14 mm long, usually in pairs, or by abortion of phyllodes in terminal racemes of 8–25 flowers, 5–8 cm long. Flowering irregularly but mainly August–December.
Pods	Fawn, smooth, papery, flat, almost straight 3–8 cm × 7–12 (–16) mm with nerve-like margins, closely rounded over seeds, occasionally narrowed between them.
Seeds	Brown, often mottled, rounded, 4–6 mm × 2·5–4 mm × 2–4 mm thick, transverse in pod; short seed-stalk straight or folded before thickening into a small aril.
Identification	Pale flowers, twin spines usually at phyllode base, flat, thin, textured pods with transverse seeds. A new subspecies has been described: subspecies *arida* Pedley in *Austrobaileya*, 1 (3), 1979, with branchlets and phyllodes on mature plants covered with soft spreading hairs; tends to have broader, more prominently veined phyllodes. Found in the more arid regions of central Australia.
Comments	Useful as a windbreak or shade tree in dry area gardens; a recommended shrub for Alice Springs area; considered to be fairly salt tolerant; requires pruning after flowering to retain shape.

Acacia victoriae

84 *Acacia murrayana*

F. Muell. ex Benth.

Common name Murray's Wattle

Meaning of name Named after J. P. Murray, medical officer and plant collector with Howitt's 1861 Expedition in search of Burke and Wills.

Distribution Widespread in arid areas on mainland Australia, in deep sandy soils from central southern Qld, western plains of NSW, southern NT to WA coast from near Shark Bay to Geraldton, south-east to Norseman and into dry mostly northern areas of SA.

Habit Spreading or erect, bushy shrub or small tree 2–5 (–8) m tall with spreading crown; bark smooth, grey or brown; branchlets angular, normally with a whitish bloom.

Foliage Variable, pale green to blue-green, flat, rather thick, linear to narrow-elliptic phyllodes 5–18 cm × 1·5–7 (–9) mm with thickened yellowish margins, faint laterals, mid-nerve prominent, ending in a small, blunt, curved point; narrowing at base into a wrinkled stalk; gland small, one obscure at base, another below the point (sometimes absent).

Flowers Numerous, bright yellow balls 7–9 mm diameter, each of 30–50 flowers, on slender stalks 4–9 (–12) mm long, in open 4–10 branched racemes, sometimes growing out into a leafy shoot. Flowering August–November.

Pods Light brown, papery, oblong 5–7·5 cm × 8–10 (–12) mm with broad nerve-like margins, flat but raised over seeds, irregularly slightly constricted between them. Immature pods sometimes deep red.

Seeds Black, oval-round, flattened 4–5 mm × 3–3·5 mm, transverse in pod; seed-stalk short, slightly thickened and folded 2 or 3 times.

Identification Phyllode venation, position and number of glands, short, open racemes of flowers, papery, broad pods with transverse seeds.

Comments A large shrub or small tree which flowers prolifically and prefers a well drained, sunny position in sandy soil. It is growing successfully in Brisbane and is a recommended species for Alice Springs. In the past, this species was mistakenly named *A. leptopetala* which is an endemic WA species.

Acacia murrayana

85 *Acacia ensifolia*

Pedley

Group 10

Common name	None known.
Meaning of name	Referring to sword-like shape of phyllodes.
Distribution	Confined to western Warrego and south-west Mitchell districts mainly on Grey Range, between Thargomindah and Adavale in Qld, growing on gravelly clay or stony soil on low scarps, often with *A. petraea* Pedley and other dry country species.
Habit	Tree up to 9 m tall with spreading crown and long, narrow, often pendulous phyllodes; bark grey-brown fissured with rough scales, often branching about 30 cm above ground; branchlets angular, smooth or covered with bloom; young branchlets sometimes wine red.
Foliage	Usually pendulous, thick green to grey-green, long straight or slightly curved phyllodes 15–27 cm × 3–8 (10) mm with slightly thickened margins and prominent central nerve ending in a long fine darkened tip; tapering at base into a yellow-green thick often curved stalk 2–5 mm long; gland near base, sometimes several others along top margin.
Flowers	Dense, very large, bright yellow balls 12–13 mm diameter, each of 50–60 flowers, on thick, smooth or glaucous stalks up to 15 mm long, in pairs or racemes up to 10 cm long. Flowering August–September.
Pods	Brown-grey, smooth, covered with bloom, almost flat, thin textured, (5–) 8–13 cm × 10–18 mm, rounded over seeds alternately, margins thickened, little if at all constricted between seeds.
Seeds	Oval, transverse to oblique in pod; straight, thread-like seed-stalk folded twice at end of seed.
Identification	Related to *A. beckleri* Tindale, but it is distinct in habit, longer narrower usually pendulous phyllodes, larger pods and transverse seeds.
Comments	Not known in cultivation, but it is a spectacular ornamental tree suited to warm to hot, dry climates.

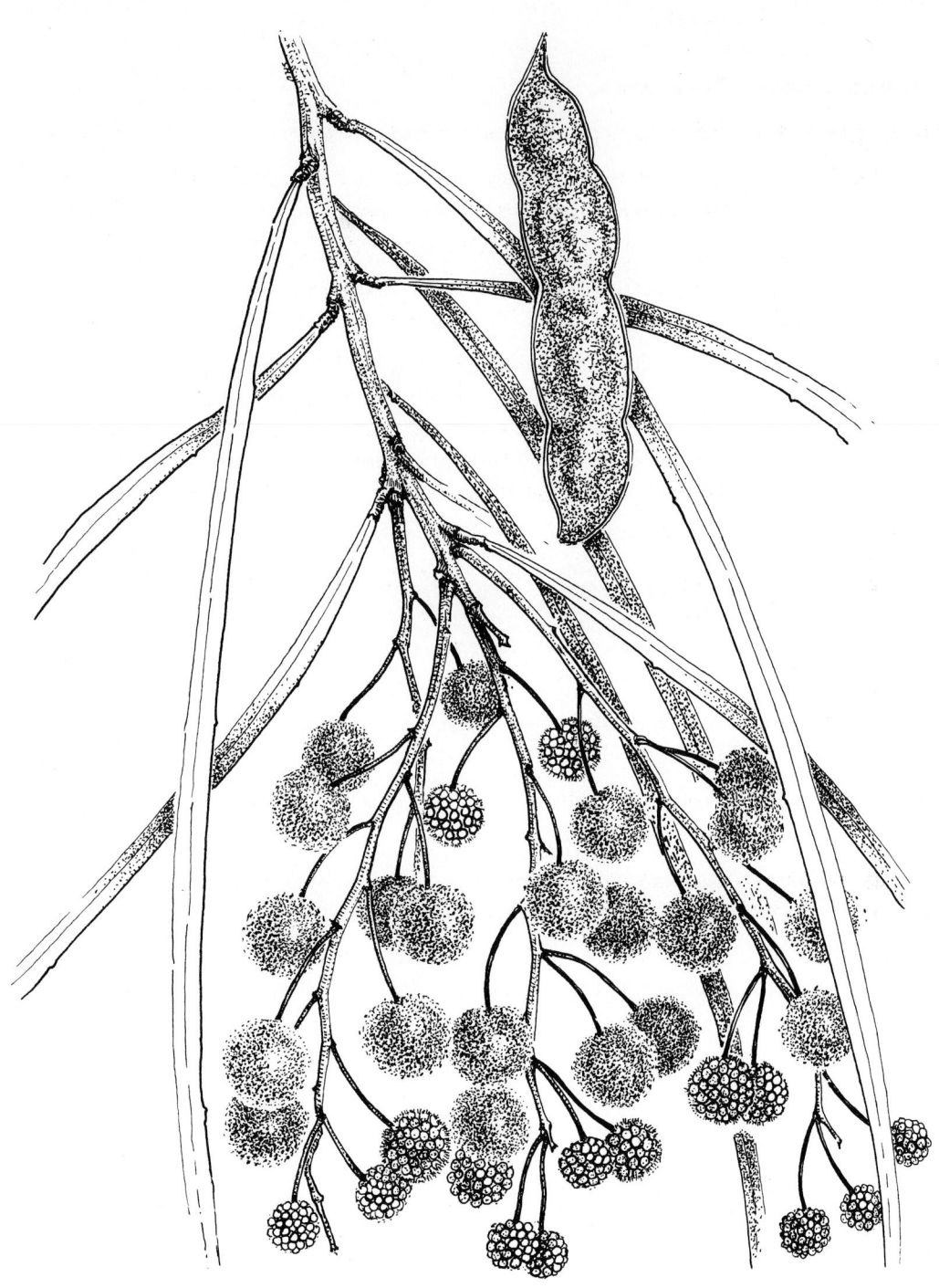

Acacia ensifolia

86 *Acacia costata*
Benth.

Group 11

Common name	None known.
Meaning of name	Referring to prominently nerved phyllodes.
Distribution	WA south-west, Stirling, Warren, Austin and Coolgardie districts, e.g. near Dandaraggan, Darling Range, Albany area.
Habit	Spreading, prickly shrub 0·5–0·6 m × 0·5 m; branches rigid, prominently ribbed, hairless or coarsely hairy, rough where phyllodes have fallen.
Foliage	Green, rigid phyllodes, spreading or slightly curving downwards, 6–14 mm × 1·5–3 mm, mid-nerve prominent, often near lower margin; margins thickened, nerve-like, tapering to a very sharp, straight point; gland basal (if present); stipules bristly, persistent. Phyllodes widest near base.
Flowers	Yellow balls 9–10 mm diameter, each of 10–20 flowers, on hairy stalks 6–10 mm long, solitary in upper axils. Flowering June–September.
Pods	Dark brown, hairy, flattened, slightly curved to c. 5 cm × 4 mm, not constricted between seeds, ending in a long curved point.
Seeds	Longitudinal in pod, c. 5 mm × 2 mm; seed-stalk ending in a cone-like aril.
Identification	Green, rigid, pungent phyllodes, widest at base, with prominent mid-nerve; large flower-heads.
Comments	Not known in cultivation.

Acacia costata

87 *Acacia colletioides*

Benth.

Common name	Wait-a-while, Spine Bush, Norka
Meaning of name	Resembling a *Colletia*, a spiny deciduous shrub from South America.
Distribution	Widely distributed through WA, SA, NSW and Vic. in open scrub or low woodland.
Habit	Spreading, often mounded, rigid, prickly shrub 1–4 m × 3–4 m with grey to reddish-brown bark, sometimes rough at base; branches round, somewhat lined, rough where phyllodes have fallen. Sometimes it forms dense, impenetrable thickets.
Foliage	Dull green, spreading, sometimes almost horizontally, round, rigid phyllodes 1–2 (–4) cm × 1–1·5 mm, with about 8 fine longitudinal nerves, tapering to a fine, very sharp point; at base sitting on a 'shoulder' where phyllode joins stem; small gland at base.
Flowers	Numerous, very bright yellow balls 4–5 mm diameter, each of 12–15 flowers, on hairless stalks 2–3 mm long, singly, in pairs or clusters of 3–5 flower heads in upper axils. Flowering July–October.
Pods	Reddish-brown, longitudinally veined, flattened, loosely curled or twisted 3–6 (–8) cm × 3–5 mm, slightly raised over and constricted between seeds.
Seeds	Black, oval, 3 mm × 2 mm, longitudinal in pod; seed-stalk fine, folded and thickened into a large cap-like yellow aril.
Identification	Pungent, 8-nerved phyllodes; the 'shoulder' on which they sit; position of gland. It is closely related to *A. nyssophylla* which once was considered a variety and which has 16 or more very fine nerves and broader, firmer pods.
Comments	Hardy shrub which is long-lived under favourable conditions. It appears to be a slow grower or unsuccessful in southern Vic.

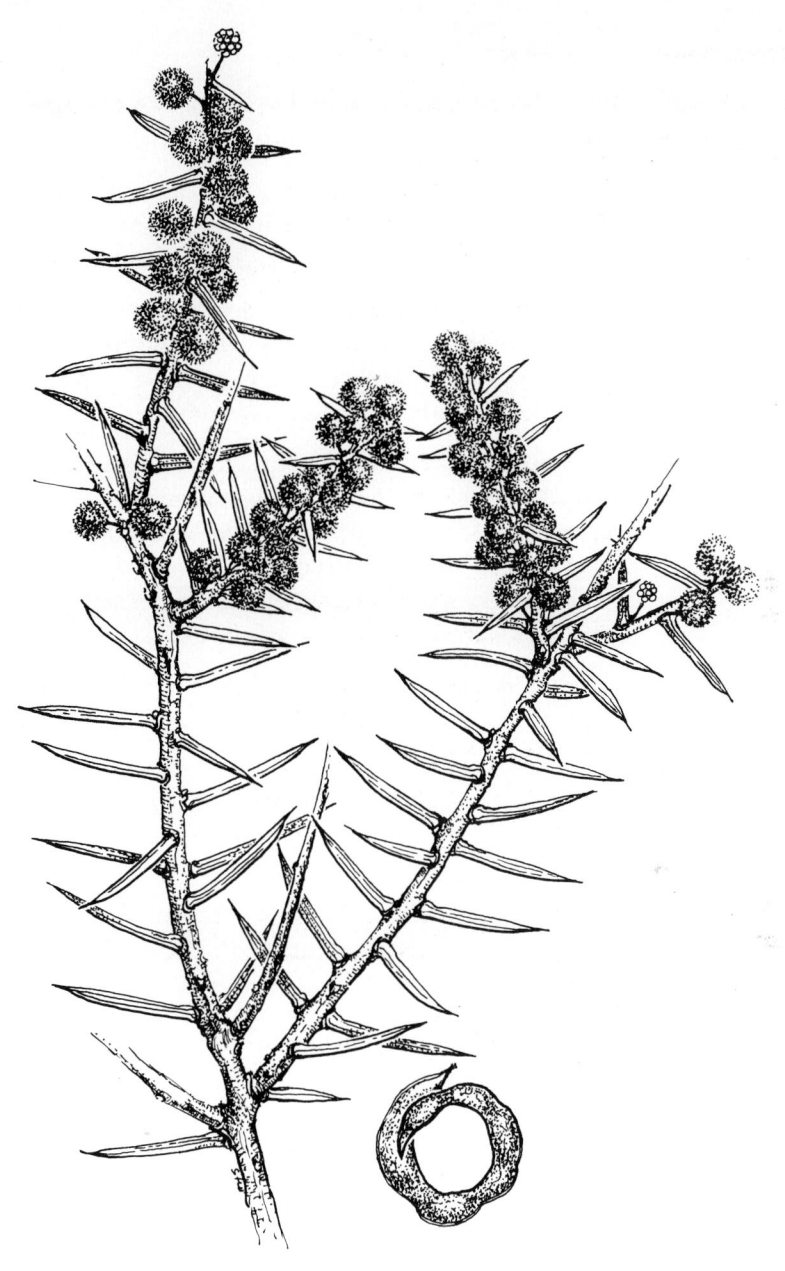

Acacia colletioides

88 *Acacia roycei*

Maslin

Common name	None known.
Meaning of name	Commemorates R. D. Royce, Curator of WA Herbarium from 1960–1974.
Distribution	WA in Irwin and Austin districts from about 90 km north of Murchison River to near Shark Bay, often growing with *A. ramulosa* and *A. wiseana* and *Casuarina* species, on sand or light loam.
Habit	Dense shrub or small tree 1·6–3·3 m × 2–4 m with fissured grey bark on main trunks, smooth on many spreading, erect branches; branchlets red-brown, round, faintly ribbed. New shoots with dense, appressed, pale yellow hairs, becoming smooth with age.
Foliage	Green to grey-green, straight or slightly curved, round, needle-like phyllodes 4–7·5 cm × *c.* 1 mm with many fine, longitudinal nerves, narrowing into a fine, brown, straight, very sharp point up to 2·5 mm long; gland, often obscure, on upper surface about 7 mm from base (occasionally a second gland above middle); stipules sharp, becoming deciduous with age, leaving woody bases.
Flowers	Fragrant, dense, bright yellow balls 6–8 mm diameter, each of 55–75 flowers, on ± hairy stalks 6–12 mm long, singly or in pairs in upper axils. Flowering August–October.
Pods	Light brown, rough surfaced, thin, curled 2·5–5 cm × 3–5 mm, flat, rounded over seeds, margins thickened, irregularly constricted, tapering at both ends.
Seeds	Longitudinal in pod; seed-stalk very short, abruptly widened into a large, bright yellow, oblique aril.
Identification	Distinguished by its stiff, needle-like phyllodes with long fine point and flower heads of over 50 flowers.
Comments	Not known in cultivation.

Acacia roycei

89 *Acacia sclerophylla*

Lindl.

Common name	Hard-leaf Wattle
Meaning of name	With hard stiff phyllodes.
Distribution	Widespread in dry areas of southern mainland especially in SA. Confined in Vic. to north-west mallee areas, in NSW to south-west plains and in WA to eastern areas.
Habit	Dense, widely spreading, often mounded, bright green shrub 1–2 m × 1–5 m with many smooth, angular, slightly ribbed grey stems, sometimes scurfy. Young growth slightly viscid.
Foliage	Phyllodes bright green, narrow, stiff, straight or curved 1–4 cm × 2–4 mm, broadest above centre, with many fine longitudinal nerves 2–5 prominent, ending in a straight or curved blunt tip; narrowing abruptly at base into a tiny light green stalk; small gland at base, if present.
Flowers	Numerous small, very bright yellow perfumed balls 4–6 mm diameter, each of 12–20 flowers, on very short smooth or hairy stalks 2–4 mm long, singly but usually in pairs, sometimes crowded in clusters. Flowering July–October.
Pods	Brown, curled and twisted, 3–6 cm × 2–4 mm, rounded over, slightly or much constricted between seeds.
Seeds	Oval-oblong, black, longitudinal in pod; short seed-stalk folded and thickened into a small cup-shaped aril.
Identification	Resembles *A. farinosa* Lindl. which differs in having hoary or white mealy branchlets and flower-stalks. Two forms are recognised: var. *sclerophylla* has striate phyllodes with prominent nerves and var. *lyssophylla* has smooth phyllodes with fine nerves (seen only under lens). *A. lineolata* from WA may prove to be the same as var. *lissophylla*.
Comments	Hardy, drought resistant shrub useful for coastal and inland conditions; requires light soil, full sun and good drainage; grown from seeds.

Acacia sclerophylla and var. *lissophylla*

90 *Acacia phlebocarpa*

F. Muell. ex Benth.

Common name	Table-top Wattle
Meaning of name	Referring to the veined pods.
Distribution	Ranging across tropical Australia from Kimberley and Ord districts of WA through NT into Qld around Mt Isa, often in low open woodland, on shallow, stony and sandy soils.
Habit	Sticky, stiff, spreading, flat-topped shrub 0·5−1·5 m tall with grey bark; branchlets sticky, with corky ribs, hairless or with hairs embedded in resin. New growth reddish, very sticky.
Foliage	Green, sticky, stiff, linear to narrow-elliptic phyllodes 2−5·5 cm × 2−9 mm, with tiny glandular dots, hairless or with sparsely hairy margins, about 12 fine longitudinal nerves, one more prominent, ending in a long, curved, non sharp point; narrowed at base; small basal gland.
Flowers	Bright yellow balls 8−10 mm diameter, each of 30−50 flowers, on sticky, solitary stalks 10−20 mm long; bracteoles tapering to a point; corolla striated. Flowering April−July, but seems to flower and fruit throughout the year.
Pods	Mid-brown, obliquely net-veined, sparsely hairy, hard, sticky, curled when dry, 5−6 cm × 4−6 mm with paler thickened margins, raised over seeds and a little constricted between them.
Seeds	Black, rounded, flattened, 3−4 mm × 3 mm, longitudinal to oblique in pod; seed-stalk with last 2 or 3 folds thickened into a cup-shaped, 2-lobed aril.
Identification	Flat-topped shape of shrub; sticky stems, phyllodes and pods, multiple veining of phyllodes; flower-heads with numerous flowers, striated corolla, sharp pointed bracteoles.
Comments	A common, conspicuous shrub around Mt Isa, it would be an attractive shrub for gardens in similar climates.

Acacia phlebocarpa

91 *Acacia cochlearis*

Group 11

(Labill.) H. Wendl.

Common name	? Rigid Wattle
Meaning of name	Spoon-shaped, referring to shape of sepals.
Distribution	WA south-west, Irwin, Avon, Darling, Stirling, Eyre and Coolgardie districts, e.g. near Eneabba (about 200 km north of Perth), Esperance area, east to nearly Eucla, on coastal sandy heaths and near coastal areas in mallee scrubs.
Habit	Rigid, prickly shrub 0·3−3 m × 0·5−2 m with smooth grey bark and angular branchlets. Young growth with scattered white hairs, especially on phyllode nerves, stems and immature pods; some are hairless.
Foliage	Green, rigid, narrow phyllodes 1·5−4 (−5) cm × 1·5−4·5 mm with three, occasionally four, prominent longitudinal nerves, tapering to a long, very sharp point, ± narrowed at base joining stem without a stalk; gland about halfway along top margin.
Flowers	Bright yellow balls 5−7 mm diameter, each of 30 or more flowers, on hairless stalks 5−6 mm long, one, two or three together in axils. Sepals spathulate, free. Flowering July−November.
Pods	Fawn-brown, usually without hairs, flattened, curled when dry, 2·5−5 cm × 2−4 mm with thickened margins; raised over seeds. Immature pods are often deep wine red.
Seeds	Dark brown, oblong 3 mm × 1·5−2 mm, longitudinal in pod; seed-stalk thickened at end into a boat-shaped aril.
Identification	Rigid, pungent phyllodes with usually three nerves, joining the stem directly, position of gland and shape of sepals.
Comments	When well-grown, a beautiful, free-flowering shrub, which has been found to become straggly in Vic. It should benefit from light pruning after flowering. Grown from cuttings and seeds.

Acacia cochlearis

92 *Acacia lanigera*

A. Cunn.

Common name	Woolly Wattle
Meaning of name	Referring to woolly hairs on phyllodes and stems.
Distribution	Widespread and frequent on stony slopes, plains and roadsides in NSW, ACT and Vic.
Habit	Rigid, spreading, often rounded shrub 1–2 m tall, sometimes prostrate, with grey, angular branchlets, usually covered with dense woolly hairs. Young growth very densely woolly-hairy.
Foliage	Green, stiff, linear to lance-shaped phyllodes 3–5 (–7) cm × 3–7 mm, usually sparsely or densely woolly hairy, rarely hairless; many prominent, longitudinal nerves, sometimes running together, tapering into a short, sharp, usually curved point; stipules, often woolly, persistent; gland sometimes present.
Flowers	Masses of large, dense, bright yellow balls, sometimes slightly elongated, 8–10 mm diameter, each of 25–30 flowers, on very short stalks, hairless or hairy, singly, in pairs or clusters in axils. Flowering August–September, sometimes earlier.
Pods	Brown, densely hairy, curled or twisted, 5–8 cm or longer × 5–6 mm, margins not thickened, rounded over and little or not constricted between seeds.
Seeds	Black, oblong 5 mm × 2–3 mm long, longitudinal in pod; seed-stalk short, folded several times, thickening into a yellow to white aril.
Identification	Venation of stiff phyllodes, usually very short-stalked flower-heads, densely hairy pods.
Comments	Hardy shrub, widely grown on a variety of soils in temperate Australia; requires good drainage and warm conditions in the south. Prune after flowering to retain shape. Grown from cuttings and seeds.

Acacia lanigera

93 *Acacia oswaldii*
F. Muell.

Common name	Umbrella Wattle or Miljee
Meaning of name	Named in acknowledgment of collections made for von Mueller by Ferdinand Oswald in SA.
Distribution	Widely ranging although rarely common; found on many types of soil in arid to semi-arid inland regions of all mainland states.
Habit	Bushy, stiff-leafed shrub or small tree 1·5–5 m and taller, with rough, fine-fissured, dark bark, usually with a short trunk and slightly angular to round smooth branchlets. Young growth often silvery hairy. It is recorded as being host to a species of mistletoe.
Foliage	Variable rigid green to blue-green lance-shaped phyllodes 3–5 (–8) cm × 3–10 (–15) mm usually smooth or with occasional hairs, evenly spaced fine longitudinal nerves, ending in a straight or curved point and at base in a dull scurfy stalk; elongated gland prominent at base.
Flowers	Pale yellow perfumed balls 5–7 mm diameter, each of 5–16 flowers, on a little stalk, usually in pairs in axils of phyllodes. Flowering October–January.
Pods	Masses of long, dark brown, woody, twisted or curled pods 6–17 (–20) cm × 6–10 mm, sometimes net-veined, with thickened margins, rounded over and irregularly narrowed between seeds. Pods often remain on bush long after seeds are shed.
Seeds	Large, black, oval, flattened 7–8 mm × 5–7 mm × 2 mm thick, longitudinal in pod; short orange seed-stalk folded and thickened at end of seed into a fleshy aril.
Identification	Very short-stalked pale flower-heads, finely nerved phyllodes and twisted woody pods, persistent on bush.
Comments	Hardy, long-lived, drought resistant, fairly salt tolerant species suitable for dry inland areas; used as a subsistence fodder for stock. Timber is hard, heavy, close-grained and durable. Grown from seed.

Acacia oswaldii

94 *Acacia heteroclita*

Meisn.

Common name None known.

Meaning of name Irregular or unusual; possibly referring to the variable size of the phyllodes.

Distribution WA south-west in Eyre and Stirling districts, from near Albany to Esperance area, in sandy heaths with *A. gonophylla*, *A. subcaerulea*, *Grevillea pauciflora* and other species.

Habit An erect, green-foliaged shrub 1–1·5 m tall with slightly angular branchlets. Young shoots minutely pale yellow silky hairy.

Foliage Variable, green, stiff, linear to lance-shaped phyllodes (2–) 4–10 cm × 2–6 (–9) mm with three prominent nerves, ending abruptly in a long, normally recurved, hard point; tapering at base in a slender stalk; gland small, about 3–8 mm from base.

Flowers Dense, bright yellow balls 5–6 mm diameter, each of over 20 flowers, on often slightly hairy stalks 3–4 mm long, singly, in pairs or clusters. Flowering September–November.

Pods Dull brown, narrow, flat 5–7·5 cm × 2–4 mm with thickened margins, very slightly raised over seeds and little if at all constricted.

Seeds Dark brown, shining, elliptical, 2·5 mm × 1–1·5 mm, longitudinal in pod; last fold of seed-stalk thickened into a fleshy, pale aril.

Identification Somewhat similar to *A. cochlearis*, but differs in its longer, less rigid phyllodes which usually have a curved non-sharp tip; a distinct basal stalk.

Comments Adaptable shrub which is being grown in south-eastern states; it is reported to withstand salt-laden winds.

Acacia heteroclita

95 *Acacia elongata*
Sieber ex DC.

Common name	Swamp Wattle
Meaning of name	Lengthened, extended, referring to phyllodes.
Distribution	Widespread in sandy soils of heaths, woodlands, often in damp areas of central tablelands, slopes and coast of NSW; in Genoa River area of eastern Vic.
Habit	A slender, erect, green leafed shrub to 4 m tall – sometimes more; branchlets and young growth angular and covered with dense, silky, downy hairs.
Foliage	Green, narrow, flat, stiff phyllodes, occasionally hairy, 5–13 cm × 1·5–4 mm with usually three prominent nerves ending in a small oblique or hooked point; narrowing at base into a curved, sometimes hairy stalk; small gland near base.
Flowers	Dense, very large bright yellow balls 8–10 mm diameter, each of 30–35 flowers, on fine hairy stalks 5–10 mm long, crowded in pairs or three in the leaf axils. Flowering August–October.
Pods	Light brown, thin, straight or slightly curved 4–9 cm × 3·5–5 mm, margins lighter coloured and flat; a few scattered hairs sometimes present; rounded over seeds and a little constricted between them.
Seeds	Shining black, 3–4 mm × 2–3 mm, longitudinal in pod; seed-stalk short, last folds ending in a small flattened aril.
Identification	Usually very large flower-heads; veining of stiff phyllodes.
Comments	Widely grown and well suited to damp positions and partial shade in eastern states; will tolerate both saline and frosty conditions. Grown from seeds and cuttings.

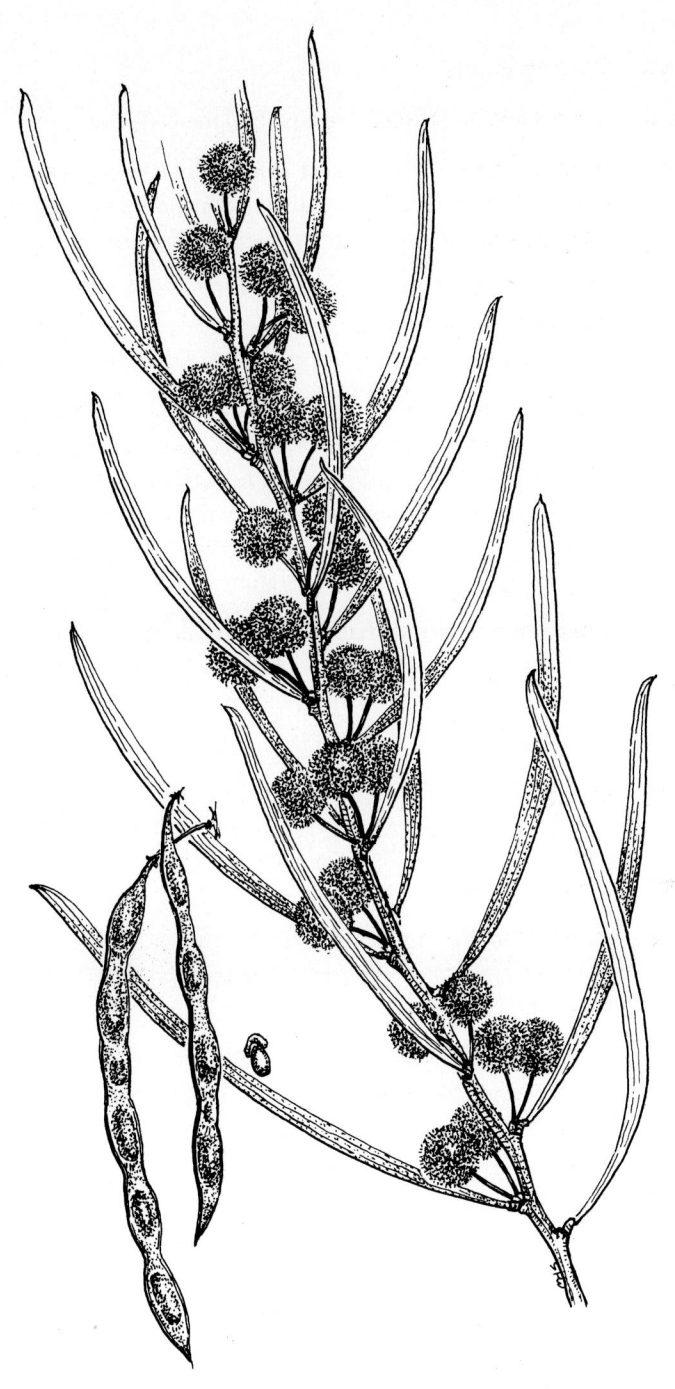

Acacia elongata

96 *Acacia simsii*

A. Cunn. ex Benth.

Common name	Sims's Wattle
Meaning of name	Named for Dr J. Sims, presumably, the collector.
Distribution	Widely distributed on sandy or gravelly soils in coastal or near coastal areas of Qld around Townsville, Atherton Tablelands, south to Proserpine; also in eastern Arnhem Land and into Papua New Guinea.
Habit	Slender, variable, smooth green-foliaged shrub 2–4·5 m tall with smooth, sometimes mottled grey bark; branchlets often reddish and angular at first. New growth light bright yellow-green.
Foliage	Green, smooth, narrow lance-shaped, usually curved phyllodes 5–12·5 (–15) cm × (2–) 5–7 mm with many fine longitudinal nerves, several more prominent in wider phyllodes; tapering at tip into a fine upturned hook and at base into a curved stalk 1–2 mm long; small basal gland with several others along top margin.
Flowers	Bright to mid yellow balls 4–7 mm diameter, each of 25–30 flowers, on fine slender smooth stalks 5–8 (−10) mm long, singly, in pairs or in greatly reduced racemes with bracts at base. Flowering at most times of the year; peak in January–February.
Pods	Brown, covered with bloom, straight or curved, flattened 5–8 cm × 4–7 mm, a little brittle when dry, with pale nerve-like margins, prominently raised alternately over and occasionally a little irregularly constricted between seeds.
Seeds	Black, ± round, compressed, 2·5–3 mm × 1·5–2 mm, longitudinal in pod; seed-stalk ± straight with last fold thickened into a club-shaped aril.
Identification	Close to *A. multisiliqua* (Benth.) Maconochie, which was once treated as a variety of the species. *A. simsii* has different phyllodes, usually longer flower stalks, and differences in seed shape and width of pods.
Comments	Grown from seeds. It requires a dry, warm, well-drained position in northern areas, where it grows quickly and often flowers in its first year. Considered a reliable shrub and highly recommended for Townsville and similar areas.

Acacia simsii

97 *Acacia leptospermoides*

Benth.

Common name	None known.
Meaning of name	Referring to the *Leptospermum* (tea-tree)–like foliage.
Distribution	WA south-west in coastal and near-coastal areas from Dirk Hartog Island in the north to Cranbrook and Lake Grace in the south.
Habit	Variable, much branched, fleshy-leafed shrub 0·5–1·5 m × 1–3 m with slender, grey branches. Young growth often covered with scattered hairs, soon becoming hairless.
Foliage	Phyllodes, variable, green, sometimes fleshy (wrinkled and flat or curved when dry) narrow-linear, obovate to ± round 4–17 mm × 1–8 mm × 1–2 mm thick; tip rounded, narrowed at base; gland on upper surface near middle in ssp. *obovata*.
Flowers	Numerous, bright yellow balls 5–7 mm diameter, each of 20–35 flowers, on hairless stalks 4–8 mm long, singly or in pairs, often crowded in upper axils. Flowering July–September.
Pods	Light brown, stiff, curled, 2–4 cm × 1–2 mm with slightly thickened margins, very slightly constricted between seeds.
Seeds	Brown, mottled, 2–2·5 mm × 1·5 mm, longitudinal in pod; seed-stalk thread-like, short, abruptly widened into a yellow-brown aril at side of seed.
Identification	Closely related to *A. ericifolia*, but differs in its smooth, shorter or broader phyllodes. Three subspecies are recognised: ssp. *psammophila* (synonym *A. psammophila*) – phyllodes small, fleshy, thick, narrowly obovate – obovate 4–7 (–9) × 1–2·5 mm (Geraldton–Mullewa district); ssp. *leptospermoides* – phyllodes linear to narrowly oblong or narrowly obovate, (7–) 9–17 × 1·5–6 mm (variable, widespread from Dirk Hartog Island to Cranbrook–Lake Grace area); ssp. *obovata* – phyllodes obovate to broadly obovate, sometimes ± round, flat and thick 5–10 (–13) × 4–8 mm (at present known only from 11–30 km north of the Murchison River).
Comments	A small shrub, occasionally grown in gardens. It is reported to tolerate clay soils, but grows best in sandy soils in full sun. Grown easily from cuttings and seeds.

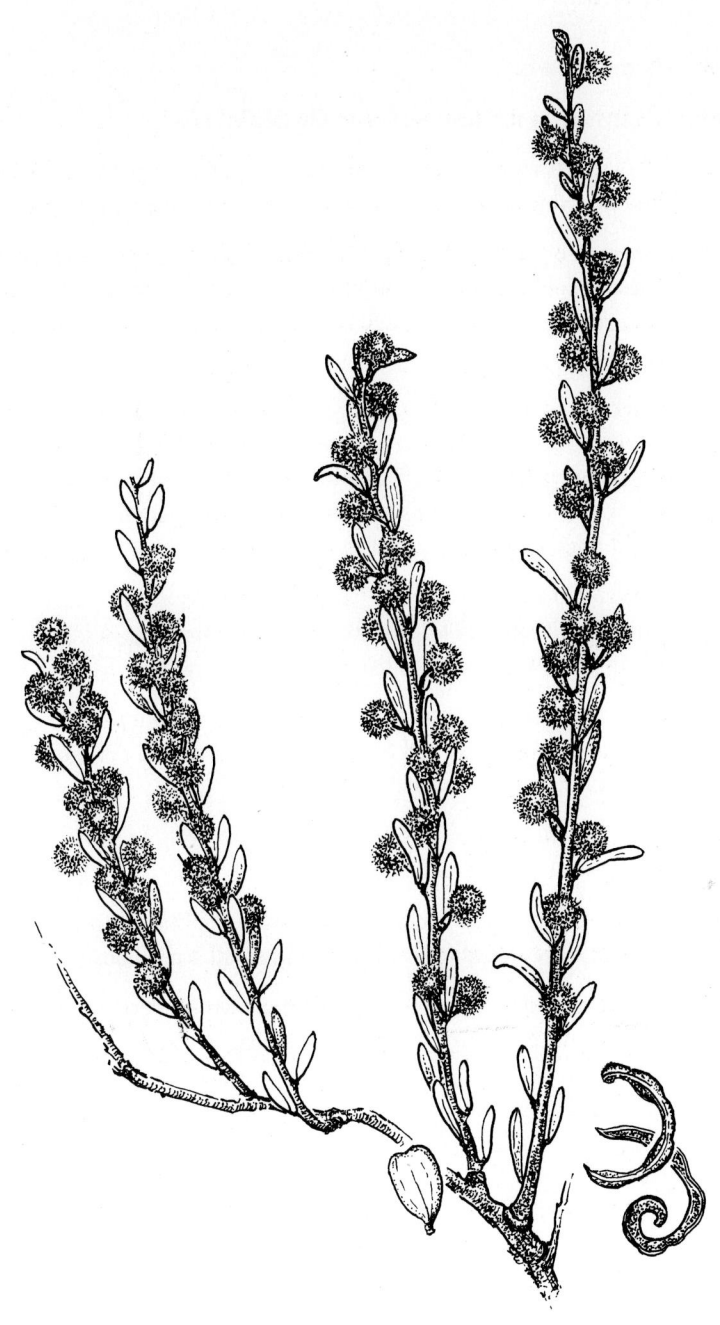

Acacia leptospermoides

98 *Acacia dictyoneura*

E. Pritzel

Common name	None known.
Meaning of name	Referring to net-like nerves of phyllodes.
Distribution	WA south-west in Stirling district, e.g. near Cape Riche, Mt Barker and Newdegate areas, on sandy loam; not a common species.
Habit	A much-branched shrub to 2 m tall with somewhat fissured bark at base; branchlets round, slightly ribbed; at tips resinous, often with a covering of very short hairs. New growth light, bright green and very sticky.
Foliage	Dark green, often sticky, stiff, wavy edged, obliquely oval to round phyllodes 0·6–3 cm × 6–20 mm, sometimes larger, with 3–6 prominent, shortly hairy, resinous, longitudinal nerves radiating from base; minor veins net-like; ending abruptly at tip in a fine, recurved point and at base in a small resinous stalk.
Flowers	Large, dense, bright yellow balls 8–10 mm diameter, each of about 30 flowers, on long, slender stalks up to 30 mm long, solitary in the axils; shining brown bracts persistent at base. Flowering March–December.
Pods	Fawn, densely hairy, leathery 2·5–3 cm × 5–6 mm with raised margins, not constricted between seeds. Pods held in clusters at the end of long, sticky stalks.
Seeds	Dull brown, oval, 3–4 mm × 2·5–3 mm, longitudinal in pod; seed-stalk short, thickened into a cap-like aril.
Identification	Distinctive, sticky, rounded, undulating phyllodes, prominently net-veined, large dense flower-heads and small hairy pods.
Comments	An attractive shrub which has been grown successfully for many years in eastern states, including Tas. It strikes reasonably well from cuttings.

Acacia dictyoneura

99 *Acacia monticola*
J. M. Black

Group 12

Common name	None known.
Meaning of name	Growing on hills or mountains.
Distribution	Widespread across northern Australia from north-west WA through NT into Qld, at Settlement Creek and in Mt Isa area, often on shallow sandy, sometimes stony soils and on ironstone.
Habit	Sticky, green-foliaged shrub or small tree 2–5 mm tall with 'minni-ritchi' bark; branchlets ribbed, sticky, clothed with moderately dense, short hairs.
Foliage	Sticky, wavy-edged, oblique, oblong-oval phyllodes (1·2–3 cm × 5–12 (−15) mm with 3–5 longitudinal nerves, semi-transparent and shortly hairy; minor veins net-like, rather coarse, ending in a small, blunt tip; narrowing at base into a small, hairy stalk; gland at base, small, disc-like; stipules small, persistent.
Flowers	Large, loose, yellow balls or rarely spikes, 9 mm diameter, each of 15–20 flowers, on hairy stalks 13–20 mm long, solitary in the upper axils; corolla thick, lined with about 5 nerves. Flowers May–July, sometimes later.
Pods	Red-brown, sticky, with raised, oblique veins sometimes hidden by matted hairs, 4–9 cm × 10–15 mm, flat, but raised over seeds, with nerve-like warty margins.
Seeds	Brown, shining, depressed in paler centre 4–6 mm × 2·5–4 mm, oblique or transverse in pod; seed-stalk folded, flattened into a conspicuous yellowish aril. Seeds collected September–October.
Identification	Distinguished by its curling, peeling bark, sticky phyllodes with 3–5 hairy, semi-transparent nerves, large flower-heads on long, hairy stalks and its lined thick corolla.
Comments	A decorative shrub suitable for growing in warm northern regions.

Acacia monticola

100 *Acacia translucens*

A. Cunn. ex Hook.

Common name	None known.
Meaning of name	Semi-transparent; presumably referring to phyllode nerves.
Distribution	WA northern districts of Kimberley and Fortescue, common in NT, often a dominant shrub in low woodland in coastal and inland areas, on sand.
Habit	Low, spreading, sometimes slender, straggling, somewhat resinous shrub 0·6–2 m × 1–3 m; branchlets smooth, dark grey, ± round or slightly angular. New growth hairy, sticky, often red.
Foliage	Variable, green, leathery, sticky, curved, mostly obliquely obovate phyllodes 1–2·5 (–3) cm × 5–10 mm, with wavy margins, several faint paler nerves, ending in an oblique or recurved, glandular tip.
Flowers	Dense, large, bright yellow balls, *c.* 10 mm diameter, each of 30–40 flowers, on stout, hairless, ribbed stalks 12–20 mm long, solitary in axils; buds large, sharp pointed. Flowering January–September, but mainly March–July.
Pods	Stalked, erect, brown, obliquely veined, ± woody 2·5–4 (–7) cm × 5–8 mm, thick, flat, with thickened margins, ending with a hooked point, narrowing to base.
Seeds	Greyish, oblong, oblique in pod; seed-stalk slightly folded, gradually thickened into a whitish cup-shaped aril.
Identification	Considered closely related to *A. nuperrima*, but differs mainly in larger flower heads with more numerous, larger flowers; usually wider phyllodes.
Comments	Long flowering shrub used extensively in public plantings in hot tropical areas.

Acacia translucens

Acacia melleodora
Pedley

Common name	None known.
Meaning of name	Referring to honey-like perfume.
Distribution	Widespread usually in sandy areas across northern Australia from inland Qld (particularly in Mitchell district) through southern NT to near eastern central WA.
Habit	Aromatic, upright, slender, usually shining bright green sticky shrub, 3–5 m × 1–4 m with angular, ribbed, warty sticky stems, sometimes covered with bloom. New growth very sticky, young phyllodes often red or yellow-green.
Foliage	Sticky, bright green, shining, occasionally dull with bloom, lance-shaped phyllodes 3–4·5 cm × (5–) 7–10 (–12) mm with prominent, yellowish nerve-like margins and net-like veins, 2–3 longitudinal nerves prominent, ending abruptly in a small straight or curved point; narrowing at base into a tiny warty stalk; prominent elongated gland near base.
Flowers	Dense very bright yellow balls 8–9 mm diameter, each of 30–40 flowers, on sticky stalks 10–20 (–25) mm long, usually singly in the axils. Flowering May–June.
Pods	Red-brown, thin, sticky, flat brittle 3–6 cm × 10–14 mm with nerve-like margins, alternately raised over seeds, little if at all constricted between them. Young pods sticky, shining green or red.
Seeds	Brown, oval 4 mm × 2·5 mm, transverse in pod; seed-stalk with one fold thickened into a small aril. Seeds mature October–November.
Identification	Similar to *A. dictyophleba* F. Muell. but distinct in Qld differing in more finely veined smaller phyllodes, smaller flowers and flower-heads. In WA and possibly SA it is considered that more investigation is required to resolve the position.
Comments	Ornamental, sweetly perfumed shrub suitable for warm, dry, northern gardens with good drainage. Grown from seeds.

Acacia melleodora

102 *Acacia ixiophylla*

A. Cunn. ex Benth.

Common name	None known.
Meaning of name	Referring to the sticky phyllodes.
Distribution	A very wide distribution ranging from southern central Qld (Miles to Jericho), drier parts of NSW (Pilliga Scrub) and in south-west WA in deep sandy soils, often growing with *Callitris* species.
Habit	Much-branched, flat-topped, spreading, usually resinous dark-green shrub 3–4 m × 2–5 m with smooth brown bark; branchlets at first angular, usually densely hairy. New growth lighter green, hairy and sticky.
Foliage	Sticky, variable but basically lance-shaped, sparsely hairy, dark-green phyllodes (1·7–) 2–4 cm × (2–) 3–7 mm, but sometimes longer and wider, with many longitudinal nerves (a few more prominent), minor net veins, usually ending abruptly in a hard curved or straight point; narrowing at base into a hairy, short, curved stalk, usually with small gland near base.
Flowers	Dense deep yellow balls 5–9 mm diameter, each of 30–40 flowers, crowded on hairy stout stalks 2–4 mm long, singly or in pairs, but usually in short racemes of 2–3 flowers; small stipules present. Flowering August–September.
Pods	Immature pods shining green, sticky, drying dull brown, curled and twisted, 3–6 cm × 2–3 mm, sparsely hairy or occasionally smooth with lighter thickened margins, slightly rounded over and irregularly constricted between seeds.
Seeds	Black, oblong, 4–5 mm × 1·8 mm, longitudinal in pod with seed-stalk thickened into a club-shaped aril and folded several times under seed.
Identification	Hairy, sticky, stiff, dark-green phyllodes with many longitudinal nerves, curled narrow pods.
Comments	Grown from cuttings or seeds. It is grown occasionally in native plant gardens and needs full sun and good drainage in southern states. The bush often is subject to attack by sooty mould.

Acacia ixiophylla

103 *Acacia dawsonii*

Group 13

R. T. Baker

Common name	Mitta Wattle in Vic.; Poverty and Dawson Wattle
Meaning of name	Named after James Dawson, surveyor, of Rylestone, NSW, from where species was first collected.
Distribution	Common on tablelands and western slopes of NSW, ACT; in a few small rocky areas near Mitta Mitta in north-eastern Vic.; very rare near Stanthorpe in south-eastern Qld.
Habit	Spindly, slender, erect shrub or small tree 1·5–4 m tall with mottled light brown bark and erect round branches; branchlets flattened with rough resinous ridges, densely hairy between ribs.
Foliage	Dull, slightly sticky, stiff, narrow lance-shaped phyllodes 3·5–10 cm × 3–9 mm with lighter margins, up to 10 longitudinal nerves, 2 rather prominent, others net-like; narrowed into a blunt or slightly recurved point; tapering at base into a brown, wrinkled leaf-stalk, about 1 mm long; indented gland near base.
Flowers	Pale or bright lemon balls 4–6 mm diameter, each of 4–6 flowers, on very short hairy stalks in small branched racemes up to 1 cm long. Viscid in bud. Flowering August–October, and as late as December.
Pods	Dull brown, smooth, thin, slightly curved or straight, 5–6 cm × 2–5 mm with lighter coloured margins, slightly constricted between seeds.
Seeds	Black, oblong, longitudinal in pod; seed-stalk threadlike then thickened under seed with one fold.
Identification	Small, very short-stalked head of few flowers in short racemes.
Comments	Suitable for cool to warm areas in full sun or partial shade in a well-drained position. Grown from seeds.

Acacia dawsonii

104 *Acacia harpophylla*

F. Muell. ex Benth.

Common name Brigalow

Meaning of name Referring to scimitar-shaped phyllodes.

Distribution Ranges widely in semi-arid regions, often in dense colonies on many soil types but commonly on fertile clay and black soils of central Qld from near coast, west as far as Hughenden and south into NSW as far as about Cobar and Nyngan in 400–700 mm annual rainfall area.

Habit Ornamental, silvery blue-green, densely crowned shrub or tree to 25 m tall with almost black, deeply furrowed 'iron' bark; branchlets slightly ribbed and young growth minutely hoary. The tree often is parasitised by several mistletoe species in Qld.

Foliage Curved, thick, sickle-shaped, smooth or slightly hairy silvery to blue-green phyllodes (7–) 10–23 (–30) cm × (5–) 10–15 (–20) mm with many faint longitudinal nerves, several more prominent with little veining between them; tapering at tip into a long thin point and at base into a long, usually curved stalk 5 mm long; raised gland usually at base.

Flowers Dense medium yellow balls 5–8 mm diameter, each of 15–30 flowers, on fine, slightly hairy stalks 15–20 mm or more long, usually in axillary clusters or in very short racemes of 6–15 flowers. Flowering is irregular, influenced by rainfall and soil moisture, and occurs mainly July–September.

Pods Reddish-brown, smooth, curved, slightly woody with raised longitudinal ridges (5–) 7–12 (–20) cm × 5–10 mm with flattened wide margins, slightly rounded over and constricted between seeds, ending in a small curved tip. Seed is not set every year.

Seeds Large, soft-coated, grey-brown, oblong, thick but flattened, 10–18 mm × 5–8 mm × 1·5–2 mm thick; longitudinal in pod with small, usually straight seed-stalk.

Identification Large sickle-shaped, grey-green, thick phyllodes, usually clustered flowers and the soft-coated seeds, which is a most unusual feature in acacias.

Comments Soft-coated seeds should be sown without treatment within a few months of collection, as viability decreases with age.

Brigalow usually is found on land which has a high agricultural potential, so is relentlessly cleared. It has the ability to regenerate strongly through suckering which ensures its survival. It is planted as a shade and shelter species in dryer, warmer climates. The timber is brown, hard, heavy and very strong, and is used as firewood, for making charcoal and on farms for fence posts, after treatment with a solution of copper-chrome-arsenic preservative.

Acacia harpophylla

105 *Acacia complanata*

A. Cunn. ex Benth.

Group 13

Common name	None known.
Meaning of name	Referring to flattened stems.
Distribution	Nowhere common and mostly restricted to sandstone areas of coastal and near coastal Qld as far north as Bundaberg and Blackdown Tableland. It extends inland to Alpha–Tambo area and south to near Grafton in NSW.
Habit	A large shrub or small slender tree to 5 m tall with smooth grey bark, downward arching branches, and smooth, conspicuously flattened, zig-zagging, often winged branchlets. New growth sometimes shining red.
Foliage	Bright green, elliptic, smooth, thick phyllodes 5–10 (–11·5) cm × (12–) 15–35 (–45) mm with 6–9 conspicuous, evenly spaced, longitudinal and fainter oblique veins, ending in a rounded tip; tapering at base into a wrinkled flattened stalk 2–3 mm long; prominent gland on top margin varying in distance from base.
Flowers	Large, deep bright yellow balls 8–10 mm diameter, each of 40–45 flowers, on smooth stalks 6–12 mm long, often in groups of 4–8 flowers (a reduced raceme). Flowering mainly December–April, but may be found in flower at other times.
Pods	Dark brown, veined, flat, sometimes covered in bloom, (5–) 10–15 cm × 6–10 mm, rounded and much raised alternately over seeds, a little contracted between each.
Seeds	Dark brown, oval, 4–5 mm × 2·5–3 mm × 2 mm thick, longitudinal or slightly oblique in pod; seed-stalk ribbon-like, wrinkled and stout completely encircling seed in a single fold.
Identification	Conspicuously flattened stems, crowded flowers. *A. homaloclada* is very similar but has usually narrower and more elongated phyllodes.
Comments	Beautiful ornamental being grown in gardens and in roadside plantings in Qld. It is also being grown successfully in Canberra.

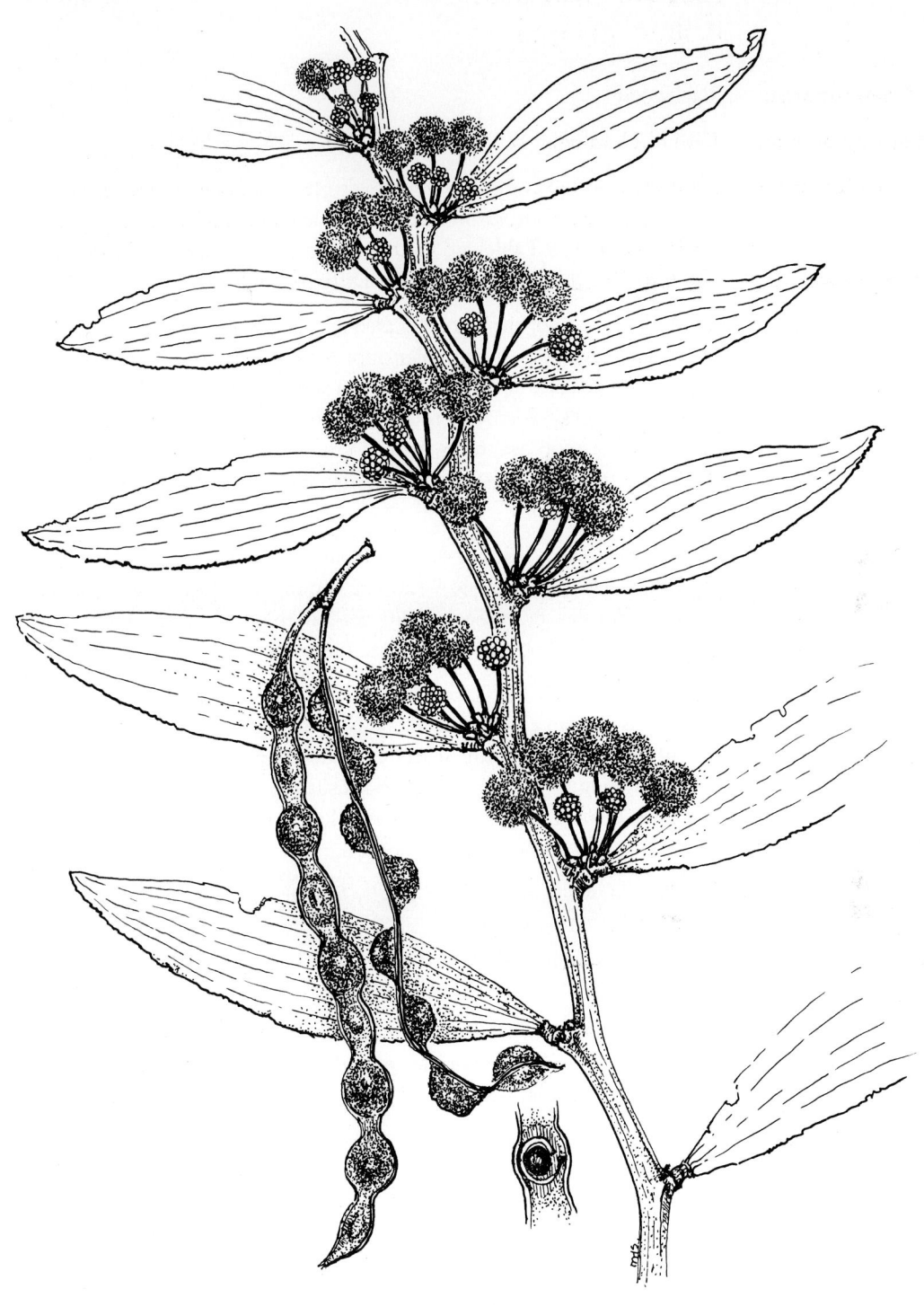

Acacia complanata

106 *Acacia melanoxylon*
R. Br.

Group 13

Common name	Blackwood
Meaning of name	With black wood.
Distribution	Widespread, often common over a wide climatic range in fertile soils in high rainfall areas, on rainforest margins and creek banks from as far north as Atherton Tableland in Qld, along the highlands of NSW and Vic. into Tas. It occurs again in Mt Lofty region of SA.
Habit	A long-lived, dull or dark green-leafed dense tree 8–30 m tall, reaching its greatest development in cooler, moister southern areas, especially in north-western Tas., usually branching a short distance from the ground; dark grey, furrowed rough bark, branchlets smooth or densely hairy, angular soon becoming round. Young growth hairy; young phyllodes often very large. Bipinnate juvenile leaves often appear after injury.
Foliage	Deep to dull green, straight or curved, smooth, oblong lance-shaped phyllodes 4–10 (–14) cm × 7–20 (–25) mm with many longitudinal nerves, 3–5 more prominent, many net-like minor nerves; ending in a small blunt point and tapering at base into a short thick stalk; gland 1–10 mm from base.
Flowers	Large dense cream to pale yellow balls 8–10 mm diameter each of 30–50 flowers, on smooth, scurfy or moderately hairy stalks 8–13 mm long, sometimes singly but usually in branched racemes of 3–5 (–8) flowers, shorter than the phyllodes. Flowering August–November; November–March in Qld.
Pods	Clusters of fawn to red-brown, usually leathery, loosely to tightly coiled, rather flat smooth pods 3–10 (–12) cm × 3·5–8 (–10) mm with thickened margins, irregularly and very slightly constricted between seeds.
Seeds	Black, oval, 3–5 mm × 1·5–3 mm, longitudinal in pod; long, crinkled, flattened pink or bright red seed-stalk, almost or completely encircling the seed in a double fold.
Identification	Resembles in general appearance *Acacia implexa* Benth. which grows in similar areas, but not in Tas. *A. melanoxylon* differs in its shorter blunt phyllodes, angular branchlets and brightly coloured seed-stalk which usually completely encircles the seed.
Comments	An adaptable species, widely grown as a specimen, shelter or shade tree in areas of sufficient rainfall. A valuable and fine timber used extensively for veneer, furniture and cabinet making. It has been cultivated since early settlement and is grown widely overseas; some commercial regeneration has been undertaken in Tasmania. The timber is strong, hard, close-grained with light reddish to dark brown heartwood and straw coloured sapwood. Grown from seeds.

Acacia melanoxylon

107 *Acacia implexa*
Benth.

Group 13

Common name	Lightwood or Hickory Wattle
Meaning of name	Entangled, referring to pods.
Distribution	Widespread, sometimes common, over a large area of eastern Australia from near Atherton Tableland in Qld, along the coast and on the tablelands south to Vic., usually in an area of about 550 mm annual rainfall.
Habit	Bushy, smooth, usually green-leafed tree to 8 m tall or taller, with rough, sometimes scaly bark and slightly angular branchlets covered with some whitish bloom. New growth light green.
Foliage	Curved, usually green, smooth, sickle-shaped phyllodes 7–18 cm × 6–16 (–25) mm with many longitudinal nerves, several more prominent, secondary nerves net-like and fine, tapering to each end; at tip into a long, fine, slightly curved point and at base into a smooth elongated stalk; gland near base sometimes inconspicuous.
Flowers	Large dense cream to pale yellow balls 10–12 mm diameter, each of 30–50 flowers, on fine smooth stalks 6–10 mm long, in short racemes or clusters of 4–8 flowers. Flowering December–March, sometimes at other times of the year.
Pods	Huge bunches of red-brown, long, slightly woody, much curved and twisted pods, 15–25 cm × 4–7 mm with lighter thickened margins, somewhat contracted between seeds.
Seeds	Shining, black, oblong-oval, 4–5 mm × 2·5–5 mm, longitudinal in pod with broad white or cream seed-stalk folded once under the seed but not encircling it.
Identification	Similar to *Acacia melanoxylon* R. Br. (Blackwood) but differs in shape of phyllodes, smoother branchlets and cream or white seed-stalk folded only under the seed, whereas in Blackwood it is bright coloured and almost completely encircles the seed in a double fold.
Comments	A fast-growing, frost resistant tree which is occasionally planted in windbreaks and for shade on slopes and tablelands in temperate areas. The timber is hard, close-grained, dark brown with yellowish lines, similar to *A. melanoxylon.* It is grown from cuttings or seeds.

Acacia implexa

108 *Acacia wardellii*

Tindale

Common name   None known.

Meaning of name   Named for Mr V. A. Wardell who drew Dr Tindale's attention to the plant.

Distribution   Restricted to shallow weathered sandstone in dry eucalypt woodlands on Thomby Range, south-east of Surat, and south of Chinchilla in Qld.

Habit   Green-leafed shrub or slender tree to 6 m tall with smooth silvery grey or white bark and smooth, ± round, ribbed branchlets, angular at first. Young tips often red.

Foliage   Large, bright green, leathery, curved, elongated oval phyllodes 10–15·5 (–17·5) cm × (18–) 20–35 (–40) mm with raised margins and netlike minor veins, two or three more prominent longitudinal nerves, the lower two running together near base, tapering at tip into a small blunt point and at base into a long stout stalk 5–10 mm long; an elongated gland at base and up to five smaller glands conspicuously raised along top margin.

Flowers   Perfumed, pale yellow balls 7–8 mm diameter, each of 30–35 flowers, on stout, scurfy, hairy stalks 2–6 mm long in stout axillary racemes up to *c.* 4 cm long. Flowering February–June.

Pods   Stalked, dark brown, wrinkled, bead-like pods 5–8 cm × 4·5–6 mm with paler nerve-like margins, rounded over seeds and deeply constricted between them, ending in a long point.

Seeds   Black, oblong, 5–6 mm × 2–3 mm, longitudinal in pod; long, thin, reddish-brown seed-stalk running the length of the seed and thickening into a boat-shaped aril. Seeds mature September–December.

Identification   Similar in some ways to *A. bancroftii* Maiden but differs in green phyllodes with several longitudinal nerves, shorter racemes of paler flowers, different pods and seeds, and habit.

Comments   Grown from cuttings and seeds. It requires a warm well-drained position and is being grown successfully as far south as Melbourne.

Acacia wardellii

109 *Acacia pendula*

A. Cunn. ex Benth.

Group 13

Common name	Myall, Weeping Myall, Boree
Meaning of name	Hanging down, referring to habit.
Distribution	Common often in pure stands on fertile heavy clay and black soils from Clermont–Emerald areas of Qld south into western plains of NSW, with an isolated occurrence near Warracknabeal in Vic.
Habit	A stately tree to 12 m tall with usually silvery pendulous foliage and branches, often weeping almost to the ground, with rough bark and slender, slightly angular branchlets which are sometimes hairy at first.
Foliage	Silvery blue-green, minutely hairy, pendulous, narrow lance-shaped straight or curved phyllodes 4·5–10 (–14) cm × 4–8 (–10) mm with several fine longitudinal nerves visible, ending in a long soft usually curved point; tapering at base into a narrow curved stalk 1–2·5 mm long; gland at or near base.
Flowers	Small pale lemon-yellow balls 6–7 mm diameter, each of 14–20 flowers, on hairy stalks 1·5–4·5 mm long arranged in short axillary racemes or clusters. Flowering irregularly, influenced by rainfall and soil moisture; usually May–June in Qld and NSW and as late as September in Vic.
Pods	Very flat, straight, sparsely hairy with usually prominently winged margins, 3–8 cm × 8–20 mm, lined with transverse net-like veins.
Seeds	Brown, oval, 4·5–7 mm × 3·5–4·5 mm, transverse in pod; seed-stalk slightly thickened and folded several times under seed. Mature seeds have been collected October–November.
Identification	Pendulous branches and foliage, small pale lemon-yellow flowers and prominently winged pods.
Comments	Beautiful specimen tree, hardy, drought resistant for dry inland areas with heavy soils and some ground moisture. It is useful as a farm shade tree, in windbreaks, for firewood, fencing and as drought stock fodder. The timber is hard, heavy, a dark rich colour beautifully closely grained and with the perfume of violets. The tree is often damaged by the larvae of the bag shelter moth. This species in Vic. is considered 'endangered'.

Acacia pendula

110 *Acacia platycarpa*
F. Muell.

Common name	Ghost Wattle
Meaning of name	Referring to broad flat pods.
Distribution	Widely ranging, sometimes common in dry inland areas from Kimberley district of north-west WA through northern part of NT to Blackall–Jericho area of Qld.
Habit	An extremely variable smooth shrub or tree to 10 m tall with roughish bark; branchlets smooth green-brown to powdery white, sometimes angular. New growth bright yellow-green.
Foliage	Variable, smooth, stiff, pale green to blue-green oblong curved phyllodes with conspicuous lighter veining, 5–12 (–20) cm × 15–35 (–45) mm with nerve-like margins and a crowded network of minor veins, three prominent longitudinal nerves, the lowest running into the lower, almost straight margin near base; rounded into a sometimes glandular, oblique blunt tip and narrowed at base into a long smooth stalk 4–8 mm long; usually several raised glands along top margin.
Flowers	Large, loose, fluffy pale yellow to cream balls 10–13 mm diameter, each of 20–30 flowers, on smooth stalks 10–12 mm long, singly, in pairs or in racemes up to 12 cm long. Flowering January–July.
Pods	Conspicuous, flat, woody, wide oblong pink-grey to brown, often covered with bloom, patterned by a network of veins, 5–13 cm × 2–3 cm, winged when young, straight and not constricted between seeds. Pods remain on bush after seeds shed.
Seeds	Dull brown, oval-oblong, flattened, 9–10 mm × 6–8 mm × 3–3·5 mm thick, closely packed transversely in pod with thick seed-stalk folded and thickened into a cap-shaped aril.
Identification	Stiff phyllodes distinguished by three longitudinal nerves and a conspicuous network of secondary veins, position of glands, large pale flowers and large flat woody pods.
Comments	Ornamental shrub grown and considered reliable in Townsville and Rockhampton. An extremely variable species which could involve more than one species but more collections and study are necessary.

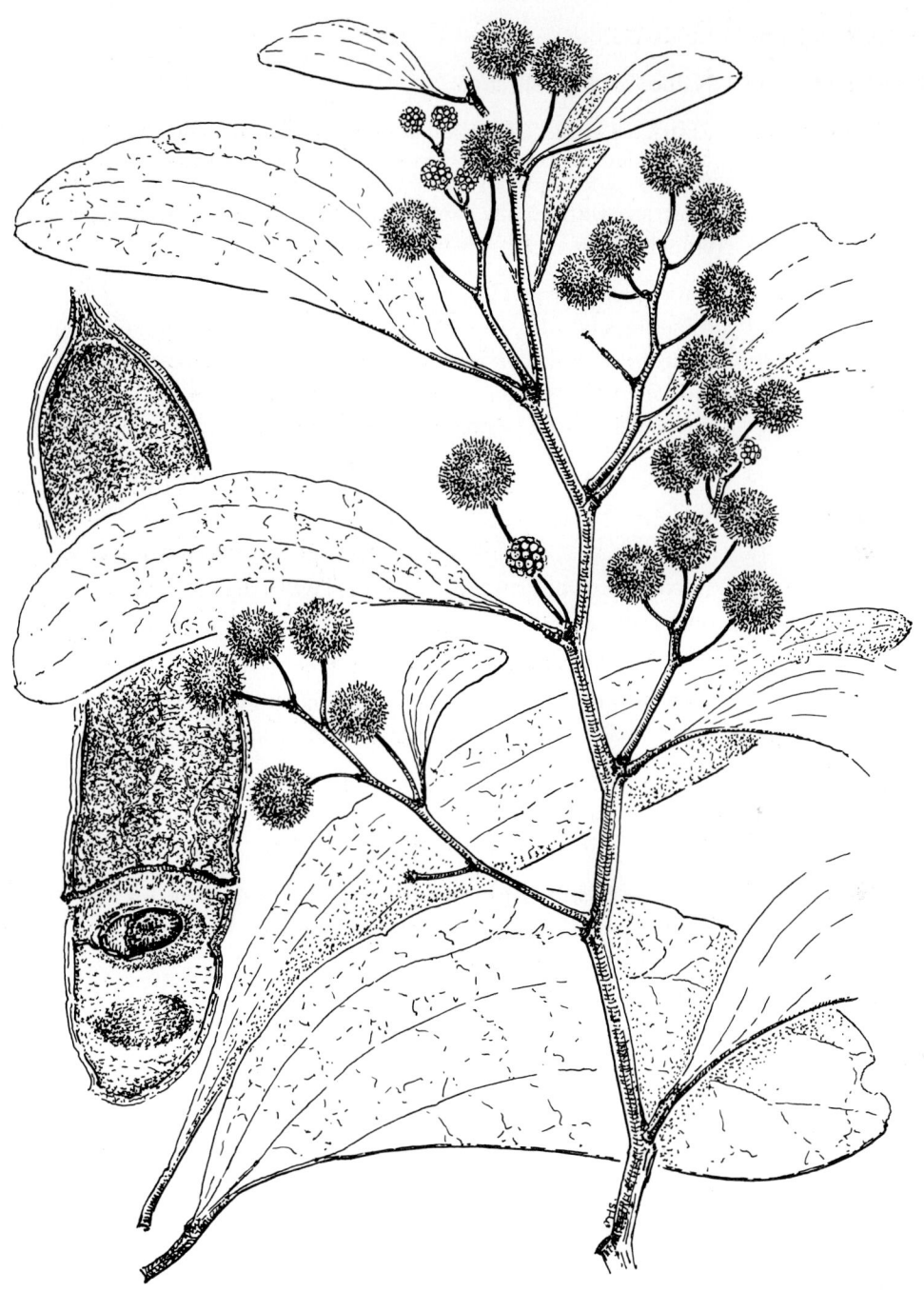

Acacia platycarpa

111 *Acacia riceana*
Henslow

Common name Rice's Wattle

Meaning of name Named for the collector.

Distribution Restricted but locally abundant in south of Tas. on moist, shady hills.

Habit A dense, green, prickly shrub or small tree 2–10 m tall, sometimes prostrate, with usually grey smooth bark and often drooping ribbed branches. Young branches reddish.

Foliage Dark green, narrow, linear, stiff phyllodes $1 \cdot 2$–$5 \cdot 5$ cm × 1–3 mm, sometimes broader, irregularly scattered or almost whorled around stem, with several longitudinal nerves, the central one more prominent, tapering into a long sharp point; narrowing at base into a tiny widened stalk; a small gland near base.

Flowers Loose pale yellow flowers in interrupted spikes $2 \cdot 5$–3 cm × 6 mm on smooth stalks 8–20 mm long, singly or several together. Flowering September–November.

Pods Narrow, brown, firm-textured, curled when dry, 4–6 cm × 2–3 mm, raised over and slightly constricted between seeds. Pods are slightly hairy when young.

Seeds Dark brown, oblong, 4–5 mm × 2 mm, longitudinal in pod; seed-stalk much folded and thickened into a fleshly aril at end of the seed.

Identification Differs from *A. axillaris* Benth. in its often weeping habit, dark green almost whorled phyllodes, loose pale flowers in spikes and in bud shape which is pointed and lopsided.

Comments It is used in street plantings and as an ornamental garden shrub in Tas. and has proved adaptable in cooler climates.

Acacia riceana

Benth.

Common name	Spur-wing Wattle
Meaning of name	Three-winged, referring to phyllode arrangement around stem.
Distribution	Widespread, in places common, sometimes in dense impenetrable thickets through the dry inland areas, western slopes and tablelands, on sand and stony gravel ridges in Qld as far north as Collinsville and Springsure, in similar areas in NSW and in a restricted area in Vic. on Warby Ranges.
Habit	Densely branched, rigid, spiny-leafed, smooth shrub or small tree 1–4 m tall spreading widely, sometimes up to 6 m across, with branches arching from ground level.
Foliage	Obliquely-triangular or narrow, curved, stiff phyllodes 2–5 cm × 2–10 mm spreading out evenly around stem, and at base running along it as a continuous wing; many fine longitudinal nerves; tapering into a long spinescent point; small gland near base.
Flowers	Numerous bright yellow loose spikes 1·5–2·5 cm long on hairless stalks 2·5–4 mm long, singly or in pairs in leaf axils. Flowering August–October.
Pods	Narrow, slightly brittle, curled or twisted 3–6 cm × 2–4 mm with nerve-like margins, slightly raised over and contracted between seeds.
Seeds	Shining black, 3–4 mm × 1·5–2 mm, longitudinal in pod; seed-stalk short, folded 3 or 4 times and thickened into a club-shaped aril.
Identification	Distinctive flat phyllodes winged down stem, and loose spike-like flowers.
Comments	Shrub suited to drier warmer areas; useful for a protective hedge as it branches from ground level. Grown from cuttings and seeds. It is considered an 'endangered' species in Vic.

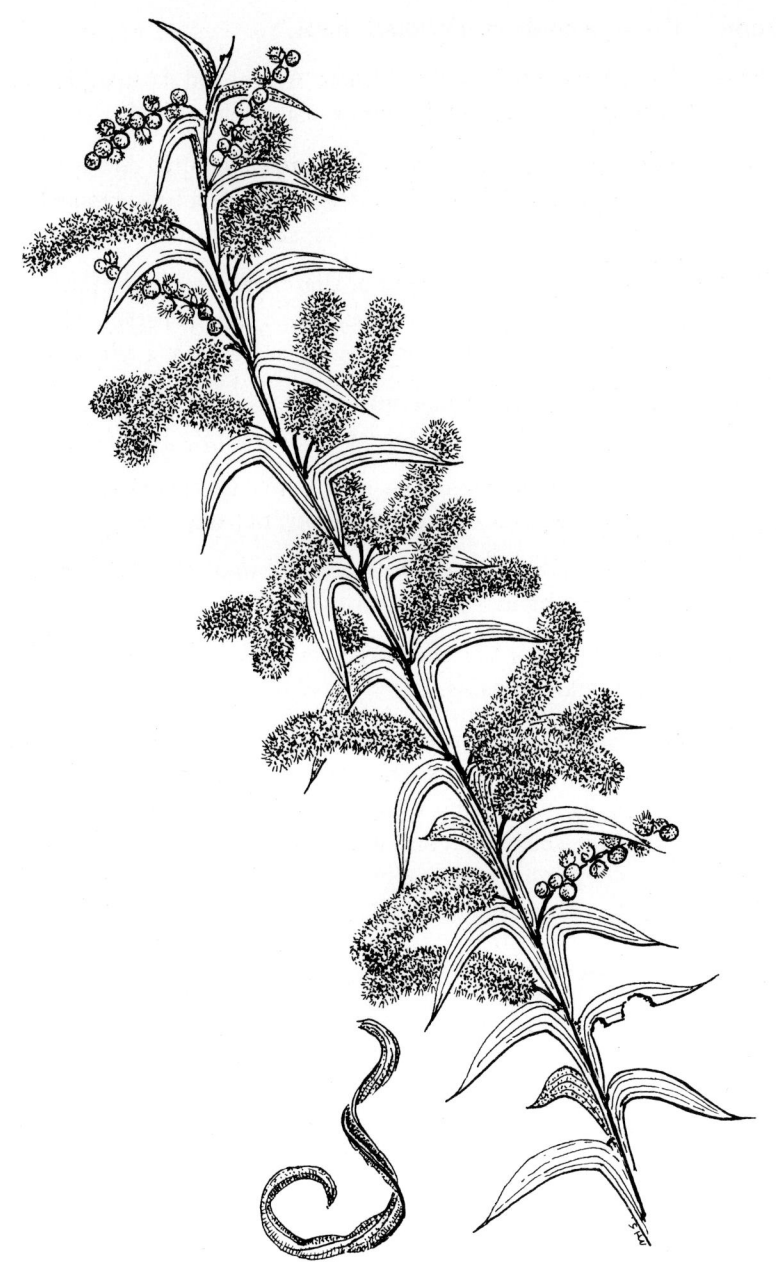

Acacia triptera

Benth.

Common name	None known.
Meaning of name	Referring to the many flower spikes.
Distribution	WA south-west in Irwin, Stirling, Austin and Avon districts, e.g. Wongan Hills, Eneabba, south to Lake Grace–Hyden areas.
Habit	Variable, many-stemmed, widely spreading, often dome-shaped shrub 0·6–3 m × 1–3 m with fissured grey bark at base; branchlets smooth, light grey, slightly angular. Young shoots covered with silky hairs.
Foliage	Usually green, stiff, straight or slightly curved, linear phyllodes 3–8·5 cm × *c.* 1 mm with prominent parallel nerves, ending in a fine, curved, non-sharp point; narrowing at base into a yellowish stalk.
Flowers	Numerous, bright yellow spikes 1–2·5 cm × 4–8 mm, stalkless, normally solitary in the upper axils. Flowering July–October.
Pods	Dull, dark brown, bead-like, thin, curled and twisted 3–8 cm × 1–2 mm raised over and evenly constricted between seeds.
Seeds	Dull black, oval, slightly flattened 3–4 mm × 2–2·5 mm, longitudinal in pod; ending in a pale cap-like aril.
Identification	Stiff, straight or slightly curved phyllodes, numerous bright yellow, stalkless flower spikes.
Comments	Densely flowered, attractive shrub which requires a warm, well-drained position. It is growing successfully in Melbourne in a frost free area, but 10 out of 11 plants died in their first winter in a frosty area, although usually it is considered frost hardy.

Acacia multispicata

114 *Acacia tenuissima*

F. Muell.

Synonyms	*A. xylocarpa* A. Cunn. ex Benth. var. (?) *tenuissima* Benth. *A. luerssenii* Domin.
Common name	None known.
Meaning of name	Extremely fine, thin – referring to phyllodes.
Distribution	Widely distributed and often common on sandy gravelly soils in eucalypt woodlands of northern central districts of Qld around Jericho north to Torrens Creek–Mt Isa region, through NT and into the northern parts of WA.
Habit	Upright, slender, slightly resinous shrub 1–4 m × *c.* 1·5 m with smooth, thin, grey to brown stems branching from ground level; smooth almost round branchlets with yellowish resinous ribs. New growth yellow-green, often sticky, covered with a dense mat of short hairs.
Foliage	Somewhat stiff, fine, smooth, upright green to yellow-green usually slightly flattened phyllodes 6–15 cm × 0·7–1 (–1·3) mm with many faint longitudinal nerves (seen under lens) occasionally with resinous margins, tapering into a stiff blunt warty tip and at base into a tiny yellow wrinkled stalk; a tiny gland near base.
Flowers	Small, dense, mid to bright yellow spikes 1–1·5 cm × 6–7 mm on sticky smooth or hairy stalks 5–10 mm long, singly or in pairs often growing out into a leafy stalk between flowers; buds often sticky. Flowering March–July, occasionally later.
Pods	Bunches of light brown, thin, curved or twisted pods 4–8 cm × 2–3 mm with thickened margins, rounded over and slightly constricted between seeds, often ending in a long thin point.
Seeds	Oval with a slight dent in centre, 2·5–3·5 mm × 1·5–2 mm, longitudinal in pod; long yellow seed-stalk folded 4–6 times to form a basal aril. Immature pods hairy.
Identification	The flowers resemble those of *A. orthocarpa* F. Muell., but it differs in the shape of the pods.
Comments	Grown from seeds.

Acacia tenuissima

115 *Acacia grasbyi*

Maiden

Group 15

Common name None known.

Meaning of name Commemorates W. C. Grasby (1859–1930) a principal of Roseworthy Agricultural College and agricultural editor of the *Western Mail* of WA.

Distribution WA, common in Austin, Ashburton and Carnegie districts, near Shark Bay, Cue and further east on muddy, stony soil and red clay.

Habit A much-branched, rounded or flat-topped shrub or small tree 3–6 m × 3–5 m with usually spreading, horizontal branches; trunk and branches (almost to the tips) are covered with rough 'minni-ritchi' bark (i.e. red and peeling off in curly, twisted flakes, see colour plate 5).

Foliage Dull green, erect, stiff, rounded phyllodes 3–6·5 cm × *c*. 1 mm with many fine longitudinal nerves (seen under lens); at tip contracted into a stiff, not sharp point. Phyllodes vary in length but are usually less than 7 cm long.

Flowers Dense, bright yellow spikes 1–2 cm × 8–10 mm on stalks 10–20 mm long, in ones, twos or threes in the upper axils; calyx large, truncate. Flowering June–September.

Pods Reddish-fawn, obliquely veined, straight or a little curved, (4–) 7–11 cm × 8–10 mm with nerve-like margins, rounded over seeds and slightly constricted between them.

Seeds Brown with paler markings, oval, flat, 8–9 mm × 7 mm, oblique or longitudinal in pod; seed-stalk white, thin, folded, ending in a small rounded aril.

Identification Sometimes confused with *A. cyperophylla*, but differs in its more rounded, spreading habit, shorter phyllodes, much longer flower stalks and larger pods.

Comments Not well known in cultivation, but is being grown successfully in Bute, SA.

Acacia grasbyi

116 *Acacia trachycarpa*

Group 15

E. Pritzel

Common name	None known.
Meaning of name	Bearing rough fruit.
Distribution	WA north-west, tropical regions of Ashburton and Fortescue districts, e.g. Roebourne, Port Hedland to Wittenoom area, in thickets on river banks and in open, stony, spinifex country.
Habit	Erect shrub or small tree 1–6 m tall with minni-ritchi bark (see colour plate 5); branchlets brown-grey, roughish, sometimes densely hairy, angular soon becoming round; tips sometimes resinous.
Foliage	Numerous, straight or curved, fine, flat phyllodes 4–9 cm × 1–2 mm, with one to three longitudinal nerves and a thickened, upper marginal nerve, ending in a fine, sharp, straight or recurved point; stipules short.
Flowers	Dense, bright yellow, slender spikes 1–2 cm × 5 mm, on hairless or hairy, somewhat resinous stalks 7–15 mm long, 1–3 in the axils. Flowering April–August.
Pods	Yellow-brown, sometimes hairy, roughly longitudinally veined, curved, flat 7–9 cm × 6–15 mm with raised paler margins, a little irregularly constricted between seeds, ending in a rounded or hooked tip.
Seeds	Yellow-brown, flat, concave, 7 mm × 4 mm, very oblique or almost longitudinal in pod; seed-stalk twisted, about 8 mm long overall.
Identification	Thin, linear, prominently 1-nerved phyllodes, slender flower spikes, broad, rough or hairy pods.
Comments	Fast growing tree used in school and street plantings in dry, tropical regions; may be suitable for stabilising sand dunes and for erosion control.

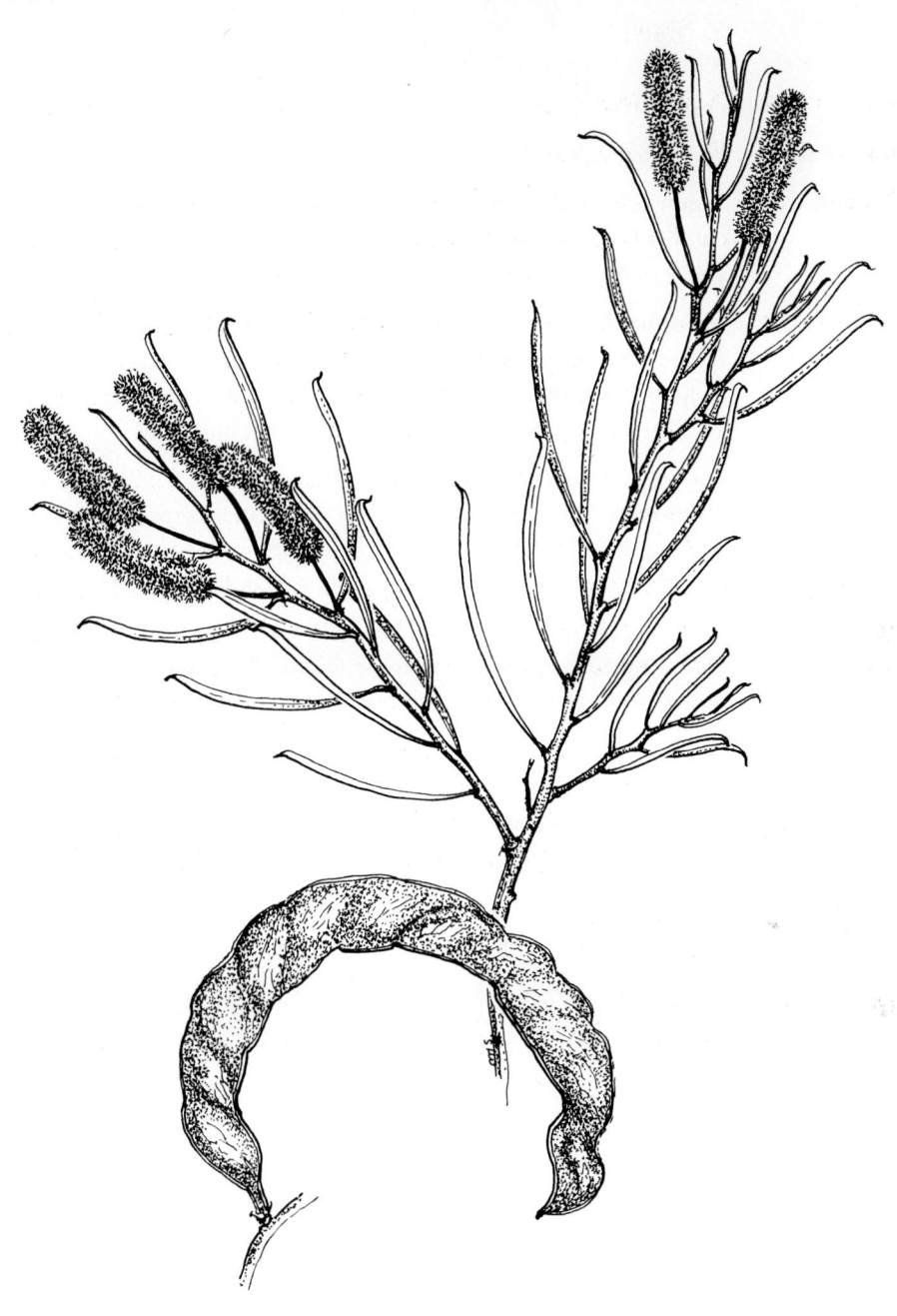

Acacia trachycarpa

117 *Acacia alpina*
F. Muell.

Common name	Alpine Wattle
Meaning of name	Of the alps.
Distribution	Confined to and often common in alpine and sub-alpine areas of Vic., e.g. Mt Buffalo, Mt Hotham, and in adjacent areas of NSW, usually at an altitude over 1500 m.
Habit	Much-branched, bushy, stiff grey-green leafed shrub, prostrate, often hugging the rocks, up to 2 m × 1–3·5 m; branchlets finely lined, angular, much flattened particularly where leaves join stem, giving a slightly zig-zagging impression. New growth purplish.
Foliage	Stiff, grey-green, obliquely oval phyllodes 1–3·5 cm × 12–17 mm, widest near top; nerve-like margins; 2 or 3 more prominent yellowish longitudinal nerves with fine net-like veins between them; rounded at tip occasionally ending in a tiny point; abruptly narrowed at base into a small flattened stalk; gland near base.
Flowers	Short, sparse, usually pale yellow spikes 1–2 cm long on short smooth stalks, singly or in pairs in the upper axils. Flowering October–November.
Pods	Dull brown, leathery, sometimes covered with bloom, longitudinally veined, slightly curved, up to 9 cm × 3–5 mm with lighter coloured margins, rounded over seeds and a little constricted between them.
Seeds	Dark brown, oval-oblong, 4–4·5 mm × 2·5 mm, longitudinal in pod; seed-stalk pale, folded several times into a cap-like aril at end of seed.
Identification	Oval shape of phyllodes, very fine net-like veins, small loose spikes of flowers and rounded stiff pods.
Comments	Useful shrub for most exposed cold situations. Grown from seeds.

Acacia alpina

118 *Acacia humifusa*

A. Cunn. ex Benth.

Common name None known.

Meaning of name Referring to the plant's often prostrate habit.

Distribution Widely distributed, usually within 100 km of the sea on shallow rocky soil inland or on sand in coastal areas of Qld from Cape York to Cape Cleveland (south of Townsville), NT, on off-shore islands and as far west as the Kimberleys in WA.

Habit Spreading, broad-leafed shrub, prostrate to 1 m tall × 1–2 m with round, usually densely hairy branchlets. New growth covered by a dense mat of hairs.

Foliage Densely hairy, broad, obliquely-oval, dull green phyllodes 4–6·5 cm × 2·5–5 cm, upper margin strongly curved and often wavy, lower ± straight; usually with three raised longitudinal nerves, the lowest running into a curved tip 2–3 mm long, the remainder curving towards upper margin with a network of minor hairy veins; narrowing abruptly at base into a grey hairy stalk 4–6 mm long; prominent circular gland at base; stipules sometimes persistent.

Flowers Dense, stumpy, bright yellow spikes 1–3 cm long on very hairy stalks 2–4 mm long, usually solitary in the upper axils; hairy concave bracteoles 2–3 mm long especially conspicuous on buds; corolla hairy. Flowering at any time of year.

Pods Upright clusters of straight, round, stiff, hairy pods 2·5–7·5 cm × 4–5 mm opening from top, with little, if any, constriction between seeds.

Seeds Black oblong, 4–5 mm × 2–2·5 mm, longitudinal in pod; seed-stalk with one or two folds thickened into a cup-shaped aril. Mature seeds may be found on bush at any time.

Identification Phyllodes are similar to those of *A. dimidiata* Benth. but long projecting bracteoles of the buds and hairy corolla separate it from that species.

Comments Grown from seeds. A shrub with distinctive foliage which would be suitable for gardens in hot, dry, rocky situations.

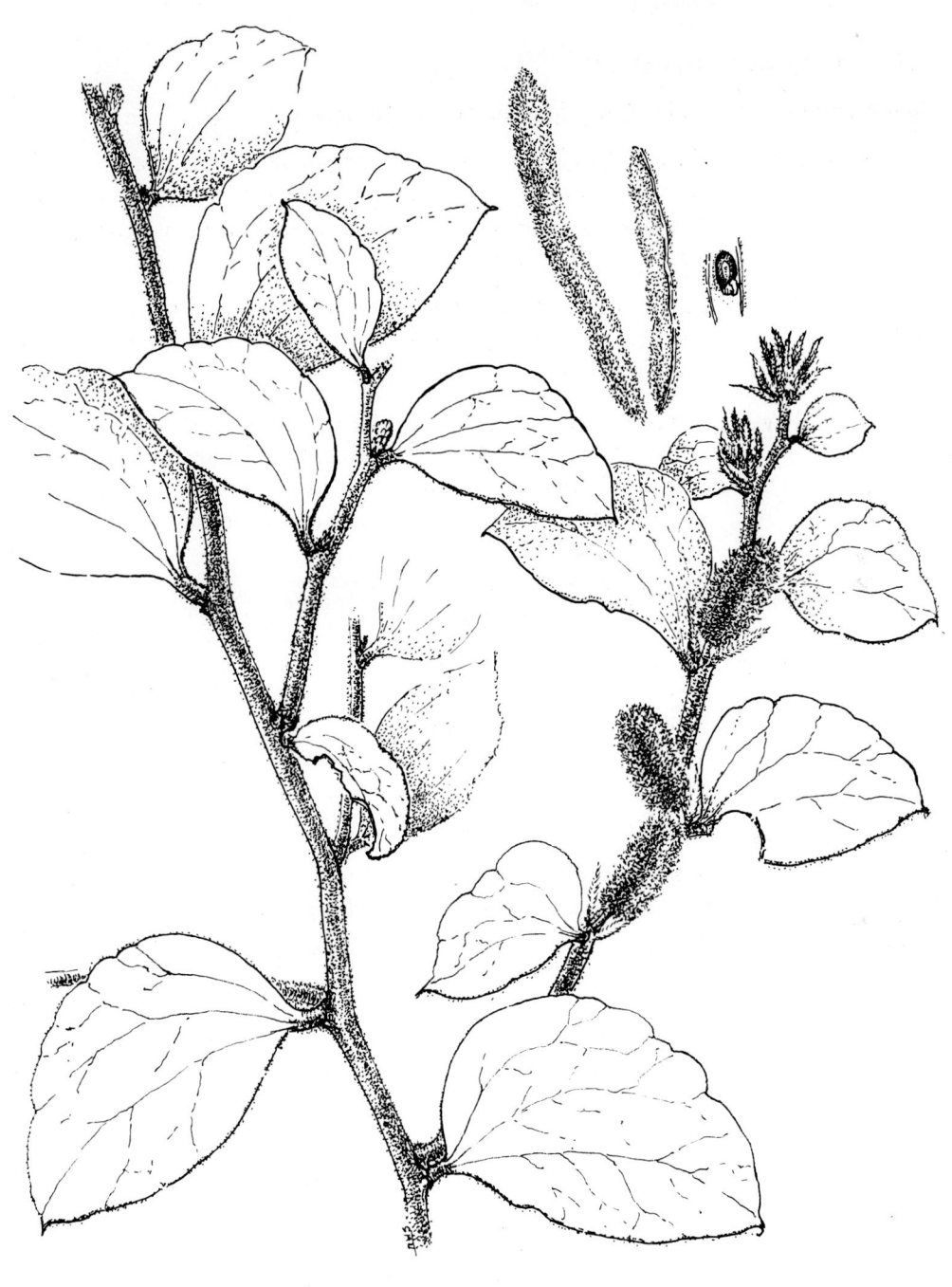

Acacia humifusa

119 *Acacia curvinervia* Group 16

Maiden

Common name	None known.
Meaning of name	Referring to curved nerves of phyllodes.
Distribution	Common on sandy soils in North and South Kennedy, Mitchell and Leichhardt districts of Qld, e.g. in Alpha, Blackall, Jericho and Pentland areas, sometimes in small stands.
Habit	Often a wide-spreading shrub or small tree 2–4 m, occasionally to 6 m tall with smooth brown bark and many stems from ground level; branchlets at first angular, scurfy or with scattered hairs. New growth scurfy and covered with dense, short hairs.
Foliage	Thick, grey-green, curved, smooth phyllodes 3·5–7 (–10) cm × 6–16 (–23) mm with a wavy top margin, many fine parallel nerves three or more prominent, ending abruptly in a short sometimes upturned warty tip; narrowing sharply at base into a short, yellowish wrinkled stalk, *c.* 2 mm long.
Flowers	Very bright yellow spikes, moderately dense 2·5–3 cm × 7 mm on stout, scurfy, hairy stalks 4–10 mm long, usually in pairs. Flowering May–June, sometimes later.
Pods	Fawn, woody, almost round, slightly curved 5–10 cm × 2·5–3 mm with conspicuous longitudinal veining; pods held upright in clusters.
Seeds	Longitudinal in pod, 3·5–4·5 mm × 1·3–1·8 mm; seed-stalk folded several times to form a short aril at end of seed.
Identification	Shape of phyllodes and curved nerves. Its nearest ally is *A. julifera* Benth., and there are some intermediates between the two species on the eastern edge of *A. curvinervia*'s range.
Comments	Grown from seeds; an ornamental shrub which would be suited to dry, hot, well-drained situations in northern areas.

Acacia curvinervia

120 *Acacia floribunda*

(Vent.) Willd.

Common name	White Sallow, Sally or Gossamer Wattle
Meaning of name	Flowering abundantly.
Distribution	Widespread in forests and woodlands of coastal NSW; common along streams in East Gippsland, Vic., and near Stanthorpe in southern Qld.
Habit	Variable bushy shrub or small tree 3–8 m tall with spreading habit and bright green foliage from ground level; grey smooth bark; branchlets angular with short appressed hairs. Young growth hairy.
Foliage	Green, smooth, thin-textured, linear to lance-shaped phyllodes 5–10 (–17) cm × 2–10 mm with fine longitudinal nerves, several prominent, sometimes sprinkled with a few scattered hairs; narrowing into a long fine point and at base into a small green stalk; gland near base, if present; secondary nerves net-like.
Flowers	Light yellow, loose spikes 2–5 (–8) cm long, produced in abundance on short, fine, often hairy stalks, singly or in pairs in the leaf axils. Flowering August–October.
Pods	Brown, longitudinally wrinkled, almost straight linear pods 6–10 (–13) cm × 3–5 mm, with a few scattered hairs; raised over and evenly constricted between seeds, ending in a long thin point.
Seeds	Black, oblong, 3–4·5 mm × 1·5–2 mm, longitudinal in pod; seed-stalk folded several times at end to form a cap-like aril.
Identification	Once included with *A. longifolia* but differs in its more bushy habit, paler more distant flowers, narrower more finely nerved phyllodes.
Comments	Widely cultivated, very hardy, suitable for cool moist conditions in many areas; will grow in clay and will withstand periods of wet conditions and frosts. Grown as a windbreak or as a quickly grown cover. Grown from seeds.

Acacia floribunda

121 *Acacia acuminata*

Benth.

Common name	Raspberry Jam Wattle, Jam Wattle (from odour of freshly cut timber)
Meaning of name	Referring to phyllodes tapering into a long point.
Distribution	WA south-west, Irwin, Avon, Darling, Eyre, Austin and Coolgardie districts; very common in low semi-arid woodlands and plains from Shark Bay area to southern coast, east to Kalgoorlie area; absent from coastal strip between Perth and Albany.
Habit	Shrub or small tree 3–5 m or taller, with a wide spreading crown and short trunk with bark fissured at base; branchlets numerous, smooth grey. Young growth usually silky with fine almost golden hairs.
Foliage	Phyllodes long, thin, sometimes curved, 6·5–15 cm × 1–5 mm, with many fine longitudinal nerves; top margin usually fringed with short white hairs; tapering at tip into a long curved point and at base into a curved stalk; gland not seen.
Flowers	Fragrant, usually dense, brilliant yellow spikes 1·3–2·5 cm × 6–7 mm, on very short stalks or stalkless in pairs in the axils. Flowering July–October.
Pods	Dark brown, linear, 5–12 cm × 5–8 mm, almost straight, ± flat or rounded over and slightly constricted between seeds.
Seeds	Dark brown-black, often shining, oblong-oval 3–4 mm × 2–4 mm, ± flattened, longitudinal in pod; seed-stalk folded two or three times and thickening into a yellowish fleshy aril.
Identification	Many-nerved long phyllodes with curved tip, top margin usually fringed with short white hairs; usually stalkless flowers. *A. acuminata* is typically a tree with a definite trunk, its phyllodes are variable in size. In the eastern part of its range its habit changes to an often dense, rounded, widely spreading shrub with narrower phyllodes (*c.* 1–1·5 mm broad). This form is known as *A. burkittii* which is found also in SA and western NSW.
Comments	Hardy species which is considered to be drought, frost, salt and lime tolerant. It is fast growing in warmer areas and is useful for shelter, erosion control. The timber is hard, heavy, durable and impervious to white ant attack. It is used as a long-lasting fencing material, occasionally for making charcoal and for craft and wood turning.

Acacia acuminata

122 *Acacia granitica*
Maiden

Group 16

Synonym	*Acacia doratoxylon* A. Cunn. var. *ovata* Maiden & Betche
Common name	None known.
Meaning of name	Granite-loving.
Distribution	Mostly confined to granite in exposed areas of New England Tableland in NSW and around Stanthorpe and Crows Nest in southern Qld, but occasionally found on sandstone. Sometimes it is the only vegetation present on the rocks.
Habit	Spreading, flat-topped, green-leafed shrub 1–1·5 m × 1·5–3 m, which may reach 3 m tall in more protected areas; branching sideways at ground level with slightly rough, lined bark on a short trunk and on base of branches; branchlets smooth, usually round.
Foliage	Stiff, usually upright, narrow elongated phyllodes (8–) 10–20 (–22) cm × (1·5–) 3–4 mm with nerve-like margins and many fine parallel veins (seen under lens), one central nerve more prominent, tapering into a stiff point and at base narrowing into a small wrinkled, curved stalk; gland, if present, near base.
Flowers	Dense bright yellow spikes (3–) 6–8 (–10) mm long on short stalks, singly or in pairs in leaf axils. Flowering August–September.
Pods	Dull mid-brown, narrow, stiff, longitudinally-lined, 4–6 cm × 2·5–3 mm with lighter nerve-like margins, evenly rounded over seeds, narrowly lengthened and constricted between them.
Seeds	Black, shining, oval, 3–3·5 mm × 1·5–2 mm, longitudinal in pod; seed-stalk pale, fine, folded 5 or 6 times to form a cap-like aril.
Identification	Related to *Acacia caroleae* Pedley, but differs in its spreading habit, stiff erect phyllodes and short almost stalkless flower spikes.
Comments	Useful dense shrub for exposed conditions; it will tolerate snow and frost but does require good drainage. Growing successfully in Tas. Grown from seeds.

Acacia granitica

123 *Acacia caroleae*

Pedley

Common name	None known.
Meaning of name	Named for Mr L. Pedley's wife, Carole.
Distribution	Widely distributed in sandy soils of Leichhardt, Maranoa and Darling Downs districts of Qld and Pilliga Scrub of NSW.
Habit	Smooth, blue-green, rounded shrub or slender tree to 6 m tall with dark grey, lightly fissured bark; with sometimes sticky or mealy angular branchlets; often with pinkish stems and dull red new growth.
Foliage	Narrow, straight, blue-green phyllodes 5–14 (–21) cm × 1·5–2·5 (–6) mm with fine crowded longitudinal nerves, central one more noticeable; margins sometimes a little glandular; tapering into a curved or straight often warty point; narrowing at base into a dull, wrinkled stalk 2–3·5 mm long; small gland at base.
Flowers	Numerous, usually dense bright yellow spikes 1–3 cm × 5–8 mm on short, scurfy, occasionally hairy stalks, singly, in pairs or often several flowers on a common stalk which may grow out into a leafy shoot. Flowering August–October.
Pods	Dull brown with slightly raised longitudinal veins, flat twisting when dry, 6–8 cm × 2–2·5 mm with thickened margins, slightly raised over and a little constricted between seeds.
Seeds	Black, 4 mm × 1·5–1·7 mm, longitudinal in pod with seed-stalk folded several times beneath seed.
Identification	Although resembling *A. doratoxylon* A. Cunn. of NSW and Vic., it is considered more closely allied to *A. burrowii* Maiden. *A. caroleae* has a different growth habit, narrower phyllodes and different floral characteristics. It was previously known as *A. doratoxylon* A. Cunn. var. *angustifolia* Maiden.
Comments	Grown from cuttings or seeds. It is under cultivation and available from native plant nurseries in Brisbane and is being tried in southern states where it requires an open, sunny, well-drained position.

Acacia caroleae

124 *Acacia doratoxylon*

A. Cunn.

Common name	Currawang, Lancewood and Spearwood
Meaning of name	Spearwood – the Aborigines used the wood for spear making.
Distribution	Widely distributed and common on central western slopes and drier parts of western plains of NSW, i.e. Pilliga Scrub near Coonabarabran; ACT; in Vic. confined to a small area of north-eastern highlands.
Habit	Grey-green leafed, large shrub or small tree to 8 m tall with dark brown, irregularly fissured bark and smooth red-brown branches; branchlets very angular at first, soon becoming round.
Foliage	Long, narrow, thick, stiff grey-green, usually slightly curved phyllodes (5–) 7–20 cm × 3–7 (–10) mm with many faint longitudinal nerves, a few running together, central one more noticeable; tapering towards point into a darkened slightly swollen flattened tip, nearly straight or recurved; small basal gland usually present.
Flowers	Dense bright yellow spikes 1·5–3·5 cm × 6–7 mm usually on smooth, short, stoutish stalks, several together on a short raceme or up to 4 flower-heads on leafy shoots. Flowering August–October.
Pods	Dull brown, very narrow, 7–10 cm × 2–3·5 mm tapering to point, rounded over and slightly constricted between seeds.
Seeds	Black, shining, 3–5 mm × 2 mm, longitudinal in pod; seed-stalk folded several times to form a cap-like aril at end.
Identification	Size of phyllodes, characteristics of flower-heads and habit of shrub or tree.
Comments	Frost hardy and drought resistant brightly flowered species which requires a well-drained position; grown successfully in all eastern states including Tas. Recognised as a useful timber, hard, heavy and durable, somewhat resembling that of *A. melanoxylon* but heavier and less grained. Grown from cuttings and seeds.

Acacia doratoxylon

125 *Acacia leptostachya*

Benth.

Synonym	*A. argentea* Maiden
Common name	Townsville Wattle
Meaning of name	Referring to slender flower spikes.
Distribution	Widely distributed from southern Cape York Peninsula as far south as Chinchilla–Charleville areas. Very common on sandy soils in Pentland–Torrens Creek region.
Habit	An extremely variable grey-green, small shrub or rounded tree to 5 m tall, up to 4 m wide with usually dark grey fissured bark. Branchlets slender, angular, hoary or silvery white with yellowish scaly ribs, later becoming round. Young growth hairy, often remaining so until plant flowers.
Foliage	Stiff, straight or curved, lance-shaped grey-green phyllodes 3–9 (–12) cm × 3–10 (–15) mm with scattered or dense hairs; numerous fine, straight, longitudinal yellowish nerves, two or three more prominent tending to run together into lower, straight margin, tapering into a fine blunt point; narrowing at base into felty grey stalk 1–2 mm long; small gland near base.
Flowers	Strongly perfumed, moderately dense bright yellow spikes 2–4 cm × 6–7 mm on short smooth or slightly hairy stalks 2–4 mm long, in pairs or more on a common stalk. Flowering June–July in northern areas; August–September in southern.
Pods	Extremely variable, from smooth, flat, rounded over seeds 6 cm × 3 mm with small longitudinally placed seeds, to flat pods covered with bloom 4 cm × 9 mm raised over larger seeds, transverse in pod.
Seeds	Brown, oblong, 2·7–4 mm × 1·5–2 mm, with fine seed-stalk in first type, folded 2–3 times beneath seed, and in the latter folded about 4 times.
Identification	Very hairy young growth often remaining until flowering, hairy angular branchlets, veining of phyllodes and seed arrangement.
Comments	Variation in pods of so great a degree is rare in one species and may mean that more than one species is included. In Townsville where it is a local shrub it is commonly seen in gardens where it grows very quickly and flowers in its first year. It is hardy and reliable in Rockhampton. It prefers an open situation and is better shaped and more compact in a dry site.

Acacia leptostachya

126 *Acacia calyculata*

A. Cunn. ex Benth.

Group 16

Synonym	*A. vilhelmii* Domin.
Common name	None known.
Meaning of name	With bracts resembling a calyx or cup.
Distribution	Restricted to northern Qld, sandy heaths on coasts or stony soils on hillsides in eucalypt woodlands in sub-coastal regions from Cape York in the north, south to about Townsville.
Habit	Small, open, smooth shrub 1–2·5 m × 1–2·5 m with rough, scaly, fawn bark; branchlets angular smooth or scurfy, much flattened and finely lined. New growth hoary with rusty red tips.
Foliage	Green to blue-green, smooth, straight or curved lance-shaped phyllodes (4·5–) 7–10 (–13) cm × (4–) 6–12 (–25) mm with nerve-like margins and many longitudinal veins, three more prominent, drawn into a small upturned blunt warty tip; tapering gradually to base into a curved stalk 2–4 mm long; gland near base.
Flowers	Perfumed loose to moderately dense, almost white spikes 1·5–3·5 cm × 7–8 mm on smooth stalks 3–8 mm long, singly, in pairs, occasionally more or growing out into a leafy shoot. Can be found flowering at any time.
Pods	Dull brown, woody, longitudinally lined, straight or curled, rounded pods 7–11 cm × 3–3·5 mm, widest towards and opening at hooked tip, tapering to base.
Seeds	Dull black, oblong, 4–4·5 mm × 1·7–2 mm × 1·2–2 mm thick, longitudinal in pod; long, threadlike, straight seed-stalk abruptly widened into a cap-like aril. Seeds have been collected in June.
Identification	Very flattened branchlets, almost white flowers, lined rounded pods.
Comments	Grown from seeds and suitable for growing in a low mixed shrubbery in a well drained position in northern areas, e.g. Townsville.

Acacia calyculata

127 *Acacia acradenia*

F. Muell.

Common name	None known.
Meaning of name	Referring to glandular warty tip of phyllodes.
Distribution	Widely distributed but not common across the northern part of Australia on hills and ranges east of Prairie in Qld, into NT to the Kimberleys in WA.
Habit	Often spindly, little-branched, woolly, dull shrub 1·5—5 m × up to 3 m with smooth grey bark; branchlets slightly angular, ribbed often sticky and covered with dense felty hairs. New growth is densely hairy.
Foliage	Thick, dull green, lance-shaped, somewhat sticky, hairy phyllodes 4·5—11 cm × 1—3 cm with yellow resinous margins and many fine longitudinal nerves, 3 to 5 more prominent, the lower ones running into the shorter, lower margin near base; top margin somewhat wavy; tapering to the tip into a small, warty, sometimes hairy point, and at base into a felty stalk 5—7 mm long; a small gland near the base, if present.
Flowers	Dense bright yellow spikes 2—4 cm × 6—7 mm on short felty stalks 1—3 mm long, usually in pairs in the upper axils but occasionally growing out into a leafy shoot. Flowering March—June.
Pods	Sticky, reddish-brown, felty, round, curved, 5—9 cm × 3·5—4·5 mm with raised yellow margins, not constricted between seeds.
Seeds	Shiny brown, oval, indented in centre, 3—4 mm × 1·5 mm, longitudinal in pod with cream seed-stalk thickened and folded several times into a helmet-shaped aril.
Identification	Sometimes confused with *A. umbellata* A. Cunn. ex Benth. but differs in usually densely haired branchlets and phyllodes, longer flower spikes, hairy stalks and longer pods.
Comments	Grown successfully in cultivation in Townsville, Qld.

Acacia acradenia

128 *Acacia umbellata*

A. Cunn. ex Benth.

Common name	None known.
Meaning of name	Umbrella-like; referring to the shrub's dense, low spreading habit.
Distribution	Widespread across the northern part of Australia, usually on stony, gravelly or shallow sandy soils in coastal or near coastal districts of Qld south of Cooktown, and in NT.
Habit	Spreading, densely branched, blue-green foliaged shrub to 1·5 m tall with brown peeling bark; many small, somewhat sticky, angular branchlets slightly ribbed at first. Young growth somewhat scurfy or silky with yellow-green to reddish tips.
Foliage	Smooth, thick blue-green, often curved, elliptical phyllodes 6–10 (–14) cm × (12–) 16–30 (–35) mm with thickened, often reddish, slightly undulating margins, numerous parallel longitudinal nerves, two or three more prominent, ending abruptly in a small brown warty tip and tapering at base into a dull brown stalk 3–5 (–7) mm long; small basal gland.
Flowers	Dense very bright yellow spikes 1–2 cm × 5–6 mm on short, often scurfy stalks 2–4 mm long, several on a common stalk, sometimes growing out into a leafy shoot. Flowering April–July, but can be found flowering at any time.
Pods	Grey-brown, woody, heavily longitudinally veined, rounded, straight or slightly curved 3–4·5 cm × 3–3·5 mm with prominent margins, held upright in clusters and opening from the top.
Seeds	Shining black, oval, 3·5 mm × 2 mm, longitudinal in pods; ribbon-like cream seed-stalk folded several times to form a basal aril. Mature seeds have been collected in June.
Identification	Closely allied to and sometimes confused with *A. acradenia* F. Muell. which is usually a spindly, little-branched shrub with longer flower spikes and larger pods.
Comments	Suited to warm, well-drained gardens in northern regions. Mature flowers and pods are found on bushes at the same time. Grown from seeds.

Acacia umbellata

Acacia julifera

Benth.

Common name	None known.
Meaning of name	With spike flowers.
Distribution	Widespread from south-east of Gulf of Carpentaria in Qld to Clarence River in NSW, usually confined to sandy coastal or near coastal regions on sandstone. Ssp. *gilbertensis* is restricted to seasonally waterlogged sandy soils of catchments of Norman, Gilbert and Mitchell rivers.
Habit	Smooth, usually green foliaged shrub or tree up to 10 m tall with dark fissured bark showing through red-brown; branchlets at first yellowish, slender and angular, soon becoming round. Phyllodes and new growth of young plants usually densely hairy; this persists on mature plants of ssp. *gilbertensis.*
Foliage	Firm textured, smooth, usually bright green but occasionally grey-green, sickle-shaped phyllodes 7–25 cm × 5–25 mm with nerve-like margins and many simple longitudinal nerves, three or more prominent; narrowed at both ends, at tip into a long dull point and at base into an often yellow stalk 2–3 mm long; small gland at base.
Flowers	Dense very bright yellow perfumed spikes 3–5 cm × (4–) 7–10 mm on usually hairy stalks 2–5 mm long, usually in pairs or occasionally in threes on short racemes which may grow out into a leafy shoot. Flowering May–June or later.
Pods	Light brown, straight, mostly round, woody, longitudinally lined when dry 5–10 (–15) cm × 2–5 mm, not constricted between seeds.
Seeds	Black, oval, 3–4 mm × 1·5–3 mm, longitudinal in pod with seed-stalk folded several times forming an oblique cup-shaped aril.
Identification	Rounded, stiff pods, densely hairy branchlets and juvenile phyllodes. Ssp. *gilbertensis* differs from the type in retention of dense hairs on mature plants; the phyllodes usually are 12–20 times as long as wide, flower spikes slightly smaller, pods broader and seeds larger.
Comments	A plant well suited to gardens in southern central Qld with open, well-drained conditions. Grown from seeds.

Acacia julifera and phyllode of ssp. *gilbertensis*

130 *Acacia gonoclada*

F. Muell.

Group 16

Common name	None known.
Meaning of name	Referring to the angled branchlets.
Distribution	Widely ranging through north-western Qld into NT and WA on stony soils; on sandy or loamy soils south of Charters Towers.
Habit	Slender shrub to 5 m tall with grey to red-brown smooth bark, branching near the ground; branchlets stout, flattened and distinctly four-angled, smooth, scurfy or covered with bloom, sometimes resinous. New growth bronze coloured.
Foliage	Thick, lance-shaped phyllodes 6–13 cm × (8–) 10–20 (–35) mm curved upwards with nerve-like margins and many fine longitudinal nerves, two or three more prominent, tapering at tip into a small straight or curved warty point and at base into a stout curved stalk 3–4 mm long; raised circular gland near base.
Flowers	Dense very bright yellow spikes 1–2 cm × 6–8 mm on stout sticky stalks 3–7 mm long, usually in pairs but also singly or occasionally in terminal panicles; calyx broad, densely hairy. Flowering May–July.
Pods	Clusters of dark brown, thin-textured, sticky, flat pods 3·5–6 cm × 3–4 mm with nerve-like margins, raised over and very slightly constricted between seeds.
Seeds	Shiny black, oval, 3–4 mm × 1·5–2 mm, longitudinal in pod; seed-stalk folded and thickened into a pale aril. Seeds mature about September.
Identification	Closely related to *A. cowleana* Tate, but differs in its usually shorter phyllodes with nerves closer together, its densely hairy calyx and shorter, usually sticky pods.
Comments	Grown from seeds.

Acacia gonoclada

131 *Acacia whitei*
Maiden

Group 16

Common name None known.

Meaning of name Named for Mr C. T. White (1890–1950), Queensland Government Botanist for 33 years.

Distribution Restricted to stony areas in eucalypt woodlands in Cook and North Kennedy districts of Qld, e.g. Herberton–Stannary Hills and Paluma Range areas.

Habit Usually a single-stemmed, spindly, green-leafed shrub 1–2 m tall with smooth, deep red, angular ribbed stems. New growth somewhat warty and dark coloured.

Foliage Smooth, green, stiff, straight or curved usually narrow-lance-shaped phyllodes 5–12 (–15) cm × 2·5–10 mm with lighter nerve-like margins and a number of parallel longitudinal nerves, the central one more prominent, tapering at tip into a blunt, short warty point and at base into a short dull-red stalk; a raised circular gland at base.

Flowers Dense very bright yellow spikes 1–2·5 cm × 8–10 mm on usually smooth short or no stalk, singly or in pairs in the axils. Flowers irregularly through the year; flowers have been collected in June, September and December.

Pods Usually straight, smooth, dark brown, flat, woody, 6–8 cm × 5–10 (–12) mm with raised wide pale margins, slightly constricted between seeds and raised over them. Brittle and curled simply when dry.

Seeds Oblong *c.* 5 mm × 4·5 mm, longitudinal in pod with thick seed-stalk folded once under seed to form a thickened aril.

Identification Dense, almost stalkless, spike flowers, veining of phyllodes and thick flat pods.

Comments Grown from seeds; useful as an ornamental shrub suited to hot dry conditions. It is known to flower when very small.

276

Acacia whitei

132 *Acacia holosericea*
A. Cunn. ex G. Don

Common name None known.

Meaning of name Completely covered with fine silky hairs.

Distribution Widely distributed through tropical parts of Qld, NT and WA. It occurs as far south as near Rockhampton in Qld and is found along creek banks as well as on well-drained hillsides.

Habit Silvery-foliaged shrub or small, often slender tree to 8 m tall with dark brown, somewhat rough bark; branchlets smooth or densely hairy with three acutely angled ribs. New growth often almost white with dense silky hairs.

Foliage Variable, large, silver to blue-green oblong-oval phyllodes 10–25 cm × 1·5–9·5 cm densely covered with short hairs or occasionally smooth, with three prominent longitudinal nerves running into the lower margin near base, and joined by faint penniveins; drawn into a tiny blunt sometimes glandular tip and narrowed at base into a long, stout, usually hairy stalk 5–12 mm long; gland near base.

Flowers Very bright yellow, slender spikes 3–6 cm × 4–5 mm on stout, densely hairy stalks 2–5 mm long, singly or in pairs. Flowering mainly June–August.

Pods Dense clusters of curled or coiled, smooth or hairy, veined, dark brown pods 3–6 cm × 2·5–5 mm with slightly thickened margins, narrowed and lengthened between seeds, little if at all constricted between them. Pods remain on bushes after seeds have fallen.

Seeds Black, oblong, compressed, 3–5 mm × 2 mm, longitudinal in pod; seed-stalk folded and narrowed into a bright yellow cup-shaped aril. Seeds collected August–October.

Identification Silky hairiness of shrub, strongly three-ribbed stems, veining of phyllodes and dense clusters of curly pods.

Comments A tree for more northern parts requiring a warm well-drained position. It is recommended for Alice Springs, grows well in Brisbane and is a fast grower in Townsville where it is used for screening purposes. It responds to light pruning after flowering. The seeds are reported to be food for the Red-wing Parrots (*Apromictus erythropterus*).

Acacia holosericea

133 *Acacia longispicata*
Benth.

Common name None known.

Meaning of name Referring to long flower spikes.

Distribution Widespread, scattered or in stands on usually sandy soils in Qld from near Ravenshoe in north to Kingaroy in south, usually not far from the Great Dividing Range. Dense stands occur in Taroom–Theodore area.

Habit Tree to 10 m tall with large silvery phyllodes; bark on trunk rough at base, becoming smooth; branchlets stout, widely flattened, angular, densely covered with short flat hairs, which extend to phyllodes and flower stalks.

Foliage Large blue-green, curved, sickle-shaped phyllodes 9–20 cm × 1·3–4 cm, much larger on juvenile plants; three longitudinal nerves more prominent, some branching minor veins, tapering at tip into a soft long point and at base into an often reddish hairy stalk, 6–10 mm long; large basal gland.

Flowers Dense, sometimes sparse, bright yellow spikes 5–12 cm × 7–8 mm on stout hairy stalks, 6–8 (–15) mm long, usually in pairs. Flowering July–September, earlier in north.

Pods Dark or reddish-brown with longitudinal veins, straight or curved smooth pods 3–9 cm × 2·5–4 mm regularly rounded over and slightly constricted between seeds, with lighter coloured margins.

Seeds Dark brown, oblong, 3·5–4·5 mm × 2–2·5 mm, longitudinal in pod; bright yellow seed-stalk folded many times under seed. Mature seed collected in November.

Identification Large, silvery blue-green phyllodes, very widely flattened hairy stems and very long flower spikes. One subspecies is recognised ssp. *velutina* Pedley; branchlets covered with long spreading dense hairs, shorter flower spikes; pod unknown. It has been collected from one area near Kingaroy, Qld.

Comments A tree best suited to northern areas; it is growing well in Townsville and is being tried in northern Vic. Grown from seeds.

Acacia longispicata

134 *Acacia aneura*

F. Muell. ex Benth.

Common name	Mulga
Meaning of name	Without nerves, referring to the seemingly nerveless phyllodes.
Distribution	Widely distributed, often found in pure stands in sandy, loamy soils of the dry inland, in open woodlands in all mainland states from near Shark Bay, WA, to within a few hundred kilometres of eastern coast of Qld and on western plains of NSW, mainly in 250–300 mm rainfall area.
Habit	Extremely variable with many forms recognised. Typically a silvery to grey-green, much-branched bushy shrub or small tree up to about 15 m tall with a short, sometimes twisted trunk, brown, roughly fissured bark and obliquely ascending branches; branchlets angular, covered with dense silvery hairs; young shoots brown, scaly, occasionally resinous.
Foliage	Silvery grey-green, thick, leathery, narrow lance-shaped flat to needle-like phyllodes 2–17 (–25) cm long $\times$ 0·9–8 (–12) mm with many faint parallel nerves obscured by a dense covering of short hairs; margins usually lighter, tips blunt, curved or oblique; small basal gland.
Flowers	Slender, short, dense bright yellow spikes 1–3 cm long $\times$ 5–7 mm on short scurfy stalks 3–8 mm long, singly, rarely in pairs; occasionally flowers are long pale spikes. Flowers throughout the year (but not every year) mainly April–July, especially after good rains.
Pods	Light brown, flat, oblong 2–5 cm $\times$ 7–14 mm, usually with winged margins as wide as 2 mm. Pods covered with raised, net-like veins, sometimes sticky, narrowed at base and very blunt at tip.
Seeds	Shining dark brown, oval, flat, *c.* 5 mm $\times$ 3–4 mm, oblique or transverse in pod; seed-stalk thin, short, 2 or 3 folds thickening into a small basal aril. Seed (not set every year) appears to take about 10 months to mature; collected between September–January.
Identification	Flat, winged pods, mostly transverse seeds; narrow, almost veinless, dull phyllodes.
Comments	Very drought and moderately frost resistant; a long-lived species for hot, dry, inland gardens; will tolerate a wide range of soils including clay. Of economic value in rural areas for fencing, firewood, and some varieties for stock fodder. Timber is very dark, hard and durable; it was used by Aborigines for spears, shields, etc. and today to make souvenirs.

Acacia aneura

135 *Acacia crassicarpa*

Group 16

A. Cunn. ex Benth.

Common name	Northern Wattle
Meaning of name	Referring to thick pods.
Distribution	Common in open eucalypt forest of coastal areas of Cook District (north of Townsville) in northern Qld.
Habit	Shrub or slender tree 2–20 (−30) m with rough, deeply fissured bark and angular, lined, scurfy branchlets.
Foliage	Smooth, usually grey-green, curved phyllodes 11–20 (−22) cm × 1–4·5 cm with slightly thickened margins and many yellowish longitudinal nerves, three more prominent tending to run into lower margin at base; minor nerves not branching; tapering at tip into a long, fine, soft point and at base into a long, wrinkled, round stalk 5–15 mm long; gland near base.
Flowers	Moderately dense yellow to bright yellow spikes 4–7 cm × 6–8 mm on scurfy, thick stalks 5–10 mm long, clustered in groups of 2–6 at the base of upper phyllodes. Flowering June–September.
Pods	Dull brown, oblong, woody, flat, twisted, 5–8 cm × 2–3·5 mm, slightly obliquely veined with very thick, straight margins.
Seeds	Black, oblong, transverse in pod with seed-stalk folded and thickened into a long aril under seed.
Identification	It is closely allied to *A. aulacocarpa* A. Cunn. ex Benth. which is found on wetter sites; it differs in less conspicuous veining on mature pods, larger phyllodes and longer leaf stalks.
Comments	Grown from seeds. The timber is hard, dark but well marked. Young roots were reported to have been cooked by Aborigines for food.

Acacia crassicarpa

136 *Acacia aulacocarpa*

A. Cunn. ex Benth.

Common name	Brush Ironbark, Hickory Wattle or Brown Salwood
Meaning of name	Referring to prominent grooving on pods.
Distribution	One of the most widely ranging species in Australia, extending from near Richmond River, northern NSW, on the edges of rainforests, creek banks and into open eucalypt woodlands in coastal and near coastal Qld to southern New Guinea, across NT into north-western WA.
Habit	Ashy or hoary foliaged shrub or tree to 15 m tall or more, with often spreading crown and usually dark grey patterned rough bark on older trees; branchlets slender, angular, hoary, occasionally sticky.
Foliage	Variable, ashy, grey-green, sickle-shaped phyllodes 5–15 cm × 6–30 mm with 1–3 longitudinal nerves more prominent than others, lower ones tending to run close together with the almost straight lower margin, ending at tip in a dark upward hook; tapering at base into usually curved stalk up to 7 mm long; gland near base.
Flowers	Pale to bright yellow ± dense spikes 2–5·5 (–8) cm × 5–8 mm on scurfy stalks 2–7 mm long, or in pairs at base of undeveloped shoots which sometimes reach 7–8 cm long. Flowering February–April in sub-tropical and April–June in tropical regions.
Pods	Light to dark brown, extremely hard, oblong, straight, usually twisted woody pod up to 10 cm × 1–2 cm, heavily and diagonally veined with raised, thickened margins and veins, often ending in a thick recurved point.
Seeds	Ovate, black and shiny *c.* 5–7·5 mm × 2·5 mm, oblique or transverse in pod; flat, rather broad seed-stalk folded 4 or 5 times into an aril under the seed.
Identification	Resembles *A. crassicarpa* A. Cunn. but differs from it in usually narrower, twisted, mature woody pods, smaller phyllodes with shorter leaf-stalks. One variety is recognised at present, var. *fruticosa* which is found on coastal peaks of southern Qld. It is described as a shrub to *c.* 3 m tall with narrower phyllodes and pods.
Comments	A species suited to gardens or parks in coastal or near coastal northern climates with summer rainfall. It is growing successfully in Townsville and Rockhampton.

Acacia aulacocarpa

137 *Acacia pulchella*
R. Br.

Common name Prickly Moses

Meaning of name Beautiful, small.

Distribution WA south-west; very common on coastal plains from north of Geraldton to Esperance.

Habit Variable, much-branched, prickly, ferny-leafed shrub 0·5–2 (–3) m tall; branchlets hairless or hairy, occasionally spinescent; spines, 1 or 2 per node, occasionally absent.

Foliage Leaves bipinnate, green or blue-green; 1 pair of pinnae; 2–8 (–11) pairs of leaflets, flat, narrow-oblong to oval 1–6× 0·5–2 mm; stalked gland at base of pinnae; stipules scaly.

Flowers Bright yellow balls 6–10 mm diameter, each of 10–40 (–50) flowers, on hairless or hairy stalks 2–15 (–20) mm long, in very reduced racemes of 1–3 flower-heads. Flowers May–October.

Pods Brown, hairless or hairy, usually flat, occasionally curved, narrow-oblong 1·5–5 cm × 3–5 mm with thickened, usually straight, paler margins, raised over but not constricted between seeds.

Seeds Dark grey-brown (mottled in var. *subsessilis*), oval, 2–4·5 × 1·5–3 mm, normally longitudinal in pod; seed-stalk small, thread-like, often slightly swollen and bent below a thickened, ± straight or once-folded aril.

Identification It is related to *A. lasiocarpa*, but differs in its flat leaflets. Five varieties are recognised at present:
var. *pulchella* – most variable; flower stalks minutely hairy (widespread Moora to Dunsborough and Albany).
var. *reflexa* – leaflets finely, softly hairy; pinnae reflexed; branches ± densely hairy (Coorow to near Gleneagle).
var. *glaberrima* – variable; leaflets 3–5 pairs, obovate, sub-glaucous; spines numerous, usually 2 per node; flower-stalk hairless 7–15 mm long (Murchison River to Ravensthorpe).
var. *goadbyi* – leaflets 5–8 pairs, 3–5 mm long; spines often few, 1 only per node or absent; branchlets prominently ribbed, glaucous (Boyup Brook to Albany–Esperance).
var. *subsessilis* – rare; flower stalks 2–3 mm long; flowers 10–19 per head; leaflets 2–4 pairs; seeds mottled, transverse to oblique in pod (Wyalcatchem to Ravensthorpe).

Comments Widely cultivated in light to medium, well-drained soils; responds to light pruning after flowering; grown from cuttings and seeds. It has been reported that in experimental work in WA this acacia in particular has shown an ability to resist attack by the root-rot cinnamon fungus (*Phytophthora cinnamomi*).

Acacia pulchella

138 *Acacia megacephala* Group 17
Maslin

Common name None known.

Meaning of name Referring to the large flower-heads.

Distribution WA—Irwin district; confined to a small area around Geraldton, in sand or loam

Habit Dull green, ferny-leafed, open, spreading shrub 1–2 m tall with grey-brown bark at base; branches finely ribbed, densely hairy, often pendulous; one, occasionally two spines, but often absent from some nodes.

Foliage Leaves bipinnate; main leaf-stalk very short; 1 pair of pinnae; 4–6 pairs of leaflets, usually dull green, obovate, flat, hairless, 3–6 × 1·5–3 mm, with rounded tip; gland stalked, at insertion of pinnae; stipules narrow, triangular.

Flowers Large, deep yellow balls 10–12 mm diameter, each of 80–90 flowers, on hairless, solitary stalks 15–25 mm long, in very reduced racemes; basal bracts, often deciduous. Flowering July–September.

Pods Dark brown with slight bloom, flat, thick, 2·5–5 cm × 3–4 mm with thickened, paler margins, little if at all constricted between seeds.

Seeds Brown, shining, oblong 3–4 mm × 1·5–2 mm, longitudinal in pod; seed-stalk fine, folded, gradually thickening into a yellow aril. Mature seed has been collected in early December.

Identification Related to both *A. pulchella* and *A. lasiocarpa*, but differs from both in its long flower stalks, larger heads of flowers with numerous flowers and larger leaflets.

Comments Bright, large-flowered shrub which is grown occasionally in eastern states. It requires a well-drained position in full sun. It strikes readily from cuttings.

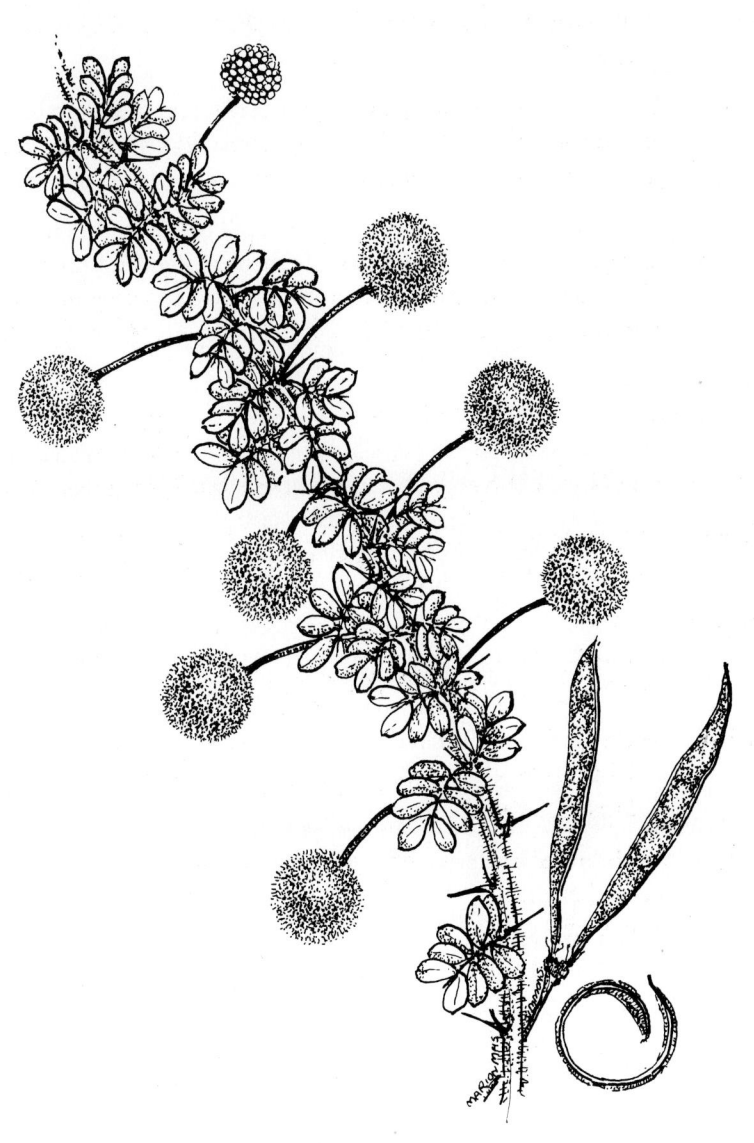

Acacia megacephala

139 *Acacia pentadenia*

Lindl.

Synonym	*A. biglandulosa* Meisn.
Common name	Karri Wattle
Meaning of name	With five glands; referring to glands which occur below each pair of pinnae.
Distribution	WA south-west in Stirling and Warren districts from Pemberton to near Albany, usually associated with karri (*Eucalyptus diversicolor*) forest, often in thickets; also found near swamps and near the coast.
Habit	Ferny-leafed, often willowy shrub or small tree 2–5 (–7) m × 2–4 m with smooth grey to brown bark; branchlets slender, prominently ribbed, hairless or occasionally with scattered hairs; axillary spines absent.
Foliage	Leaves bipinnate, dark green, mostly hairless; 2–5 (–8) pairs of pinnae, small pair nearest stem often short-lived; 2–4 pairs of leaflets on nearest pinnae; 12–30 pairs on other pinnae; leaflets usually oblong, blunt (1·5–) 3–6 × 1–2·5 mm, flat or slightly recurved, usually hairless, occasionally sparsely hairy, dark green above, light green below; prominent circular gland on upper surface of stem near insertion of each pair of pinnae, sometimes absent from smallest pair of pinnae.
Flowers	Usually large, pale yellow balls 5–10 mm diameter, each of 20–25 flowers, on long, smooth stalks 6–20 mm long, in very reduced racemes, usually clustered 6–8 in leaf axils. Flowering September–December.
Pods	Dark brown, hard, flat, narrowly oblong 1·5–5 cm × 3–4 mm with thickened yellowish margins, not constricted between seeds, slightly raised over them.
Seeds	Light brown, shining, oblong 2·5–3 × 1·5–2 mm, longitudinal in pod; seed-stalk very short, flattened and abruptly swollen into a club-shaped or folded, pale yellow aril.
Identification	Closely related to *A. subracemosa*, but differs in its usually hairless branches, leaves and flower stalks, its several glands and clustered flowers. Both grow in karri forests but their ranges do not overlap.
Comments	Tall shrub suitable for partially shaded or damp positions, growing successfully in NSW, Vic. and Tas. It is reported as being lime tolerant.

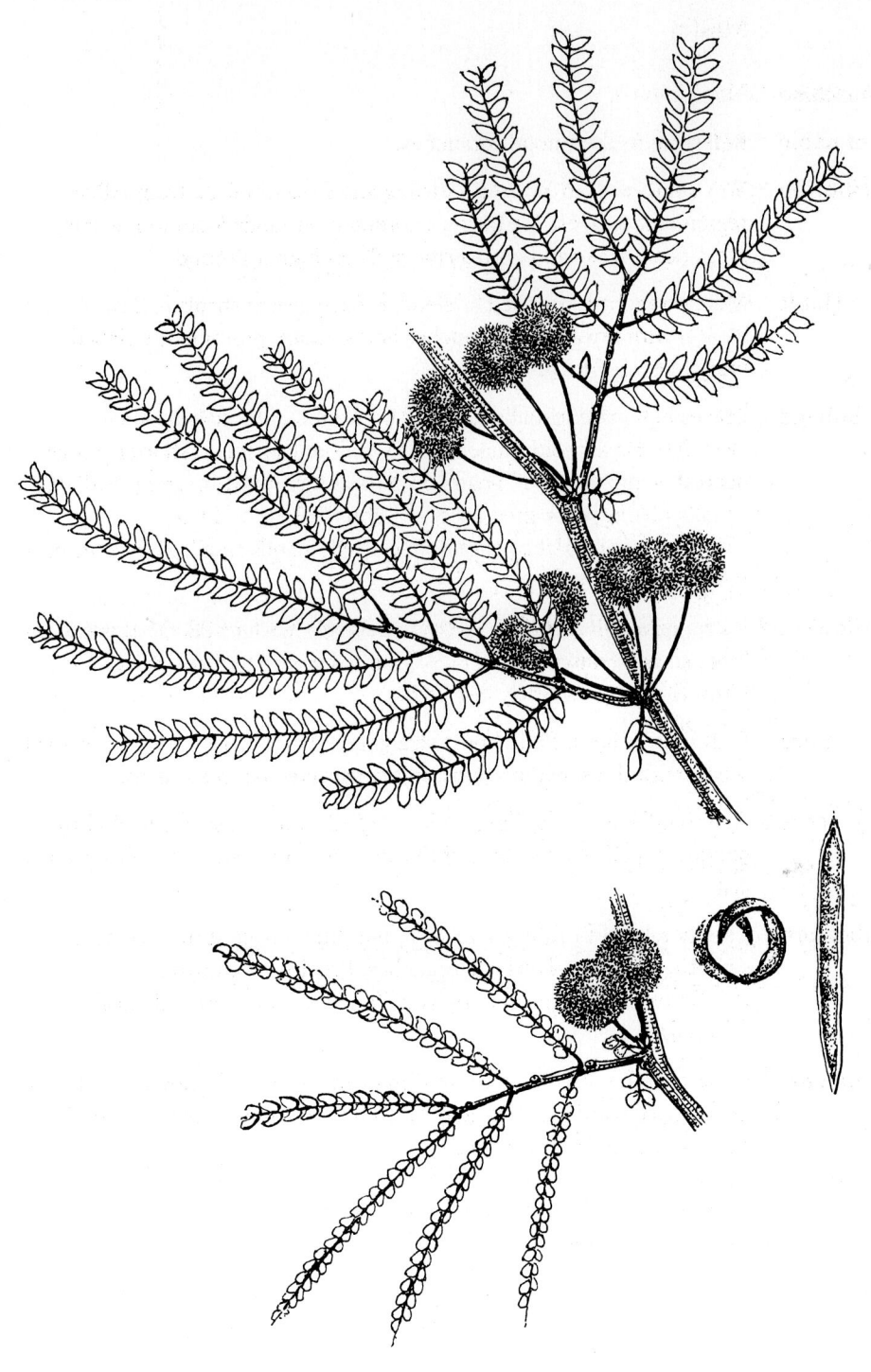

Acacia pentadenia

Acacia leioderma
Maslin

Common name None known.

Meaning of name Referring to the smooth branches.

Distribution WA south-west in Warren, Stirling and Eyre districts, from Albany region east to West Mt Barren; common on sandy loam in low-lying areas near creeks and on lateritic soils on higher ground.

Habit Much-branched, rounded or slender, ferny-leafed shrub 1–2 m × 2–3 m with brown-grey branches and smooth, prominently ribbed branchlets. Spines absent.

Foliage Leaves bipinnate, usually hairless, on prominently ribbed stems; (1–) 2 (–3) pairs of pinnae; 2 (–3) pairs of leaflets on shorter pinnae nearest stem; 4–8 (–13) pairs on pinnae furthest from stem; leaflets usually oblong, dark green above, lighter below, 5–11 × 1·5–3 (–4) mm, flat or slightly recurved; glands small, if present, near insertion of pinnae.

Flowers Large, pale yellow balls 7–10 mm diameter, each of 28–35 flowers, on long, smooth stalks 10–20 (–30) mm long, one to three in the axils. Flowering August–early November.

Pods Dull brown, hard, flat, almost straight 3–4 cm × 6–8 mm, raised over seeds, with thickened margins, not constricted between seeds.

Seeds Tan to grey-brown, oblong 2–3 × 1·5–2 mm, transverse to slightly oblique in pod; seed-stalk slightly expanded and bent below a thickened aril.

Identification Closely related to *A. empelioclada*, but differs from it in having smooth, prominently ribbed branches. Previously known as *A. nigricans* which is an entirely different shrub found only near Esperance.

Comments It has been reported to grow to 3 m or more in cultivation and is being grown successfully in Vic. and Tas. Light pruning after flowering is suggested to maintain shape.

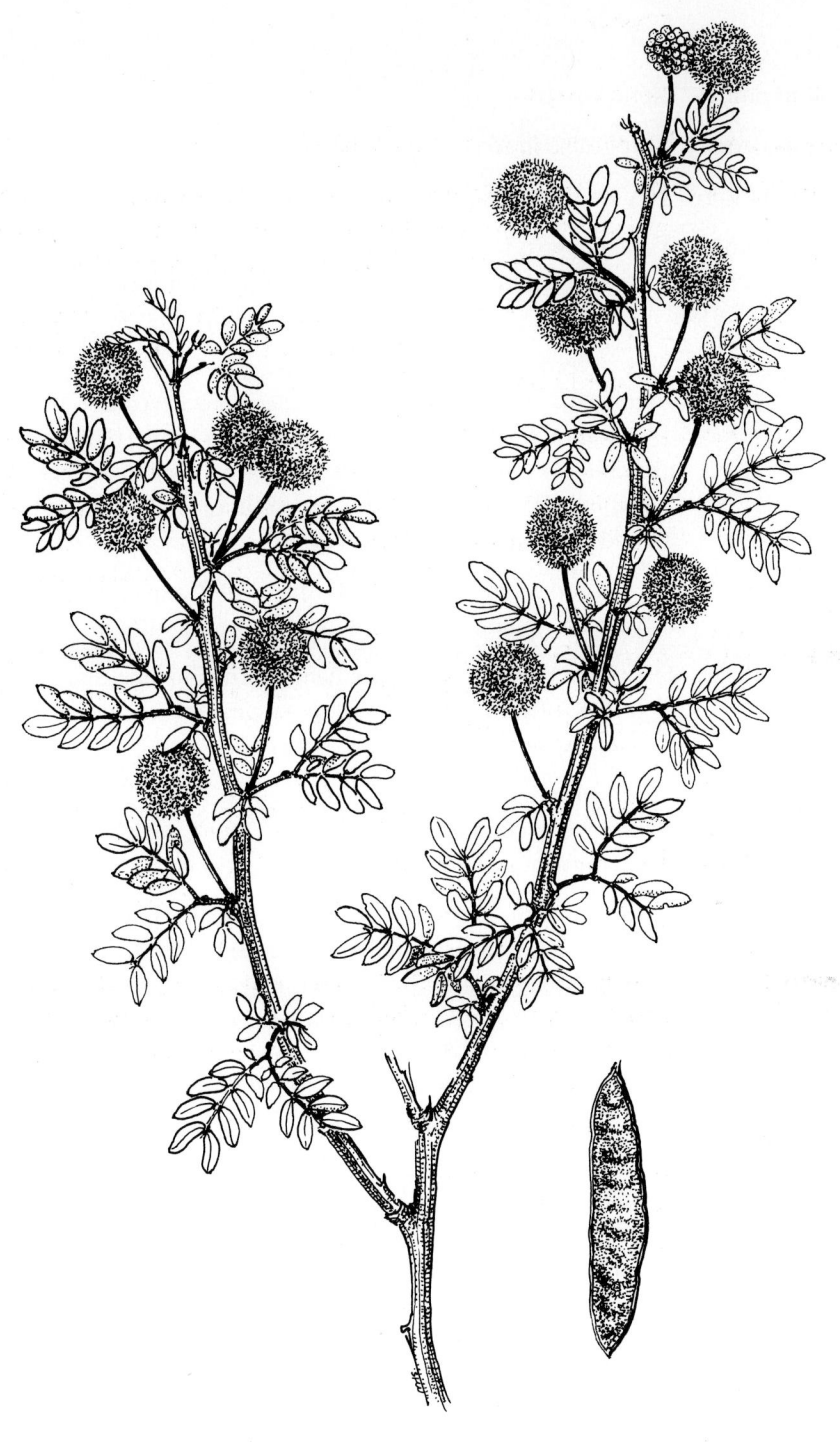

Acacia leioderma

141 *Acacia varia*

Maslin

Group 17

Common name	None known.
Meaning of name	Changeable in form.
Distribution	WA south-west districts from New Norcia to Esperance.
Habit	Variable, soft, ferny-leafed shrub 0·25–1 m tall, sometimes suckering; branchlets faintly lined, hairless to hairy; axillary spines absent.
Foliage	Leaves bipinnate, hairless or hairy; (1–) 2 (–3) pairs of pinnae; 2–3 pairs of leaflets on pinnae nearest stem; 2–5 (–7) on pinnae furthest from stem; leaflets usually oblong, flat to recurved 2–6 (–7) × 1–2·5 mm, normally dark green, nerveless above, bluish-green, prominently 1-nerved below, usually hairy; gland on upper surface below insertion of lower pinnae, sometimes another gland below second pair of pinnae (var. *parviflora*).
Flowers	Pale or dark yellow spikes 7–20 mm × 4–7 mm on solitary, usually densely hairy stalks 7–15 mm long. Flowering spring.
Pods	Dark brown to greyish-brown, hairy, narrow-oblong, flat 1·5–3·5 cm × 3–6 mm with thickened, straight margins, raised over seeds, contracting at tip into a short, sharp point.
Seeds	Dark brown-black, shining, oval to round 2–3 × 1·5–2 mm, transverse or oblique in pod; seed-stalk short, folded once before thickening into a folded, pale aril.
Identification	Related to both *A. drummondii* and *A. luteola*, but differs from former in type of hairy covering of branchlets and much recurved, softly hairy leaflets; from the latter by position of gland and construction of flowers. Three varieties are recognised: var. *varia* – flowers ± cream, stalks usually densely hairy (scattered, Gleneagle to Margaret River and Many Peaks). var. *crassinervis* – flowers deep yellow, leaflets ± flat (Williams to Katanning). var. *parviflora* – flowers deep yellow, leaflets prominently recurved; second gland often present (Arthur River to Esperance area).
Comments	Some varieties are being grown successfully in eastern states. It is considered suitable as a tub plant. Lightly prune after flowering, if necessary, to keep plant bushy.

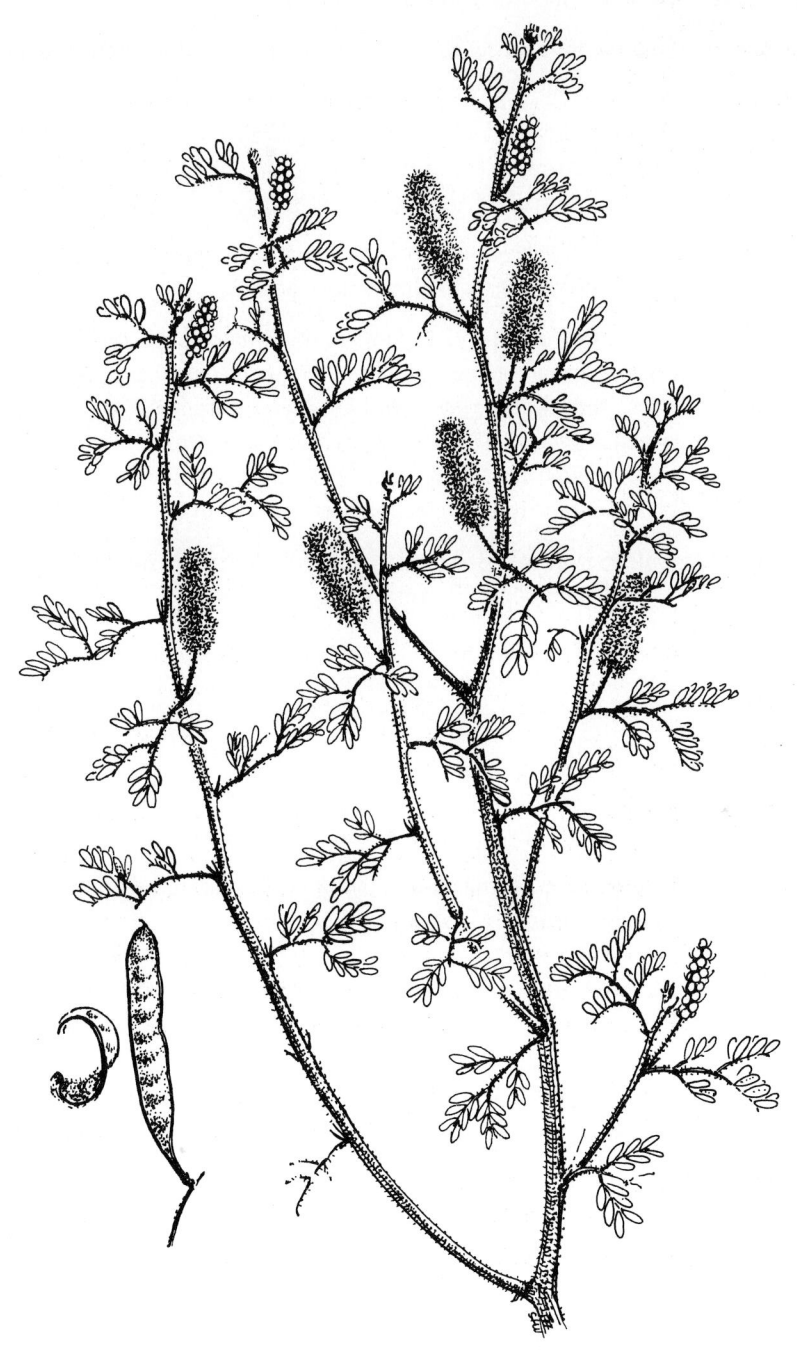

Acacia varia var. *parviflora*

142 *Acacia drummondii*

Lindl.

Common name Drummond's Wattle or Drummond's Acacia

Meaning of name Commemorates James Drummond (1784–1863), renowned botanist and collector in WA.

Distribution WA south-west, Darling, Avon, Warren, Stirling and Eyre districts, scattered from north of New Norcia to south-west of Ravensthorpe. An under-shrub of the south-west forests.

Habit Variable, ferny-leafed shrub 0·3–2 m tall; branches slender, hairy, sometimes hairless; axillary spines absent.

Foliage Bipinnate leaves; 1–2 (–3) pairs of pinnae; (1–) 2 (–4) pairs of leaflets on smaller pinnae nearest stem; 2–4 (–6) pairs on pinnae farthest from stem; leaflets green or blue-green, variable in shape and size, flat to much recurved (ssp. *affinis*), hairless or sometimes densely hairy (ssp. *affinis*); glands circular, below insertion of pinnae (sometimes gland absent from pinnae farthest from stem).

Flowers Golden spikes 1–3·5 (–4·5) cm × 5–7 mm on solitary, hairy stalks 1–3 (–4) cm long. Flowering July–October.

Pods Light, dark or greyish-brown, hairy or hairless, stiff 1·5–4 (–5) cm × 3–8 mm, with thickened margins, slightly raised over and slightly constricted between seeds; at tip, abruptly narrowed into a short point.

Seeds Light to dark brown, oval 2–3·5 mm × 1·3–2 mm, transverse to oblique in pod; seed-stalk folded and thickened into a pale yellow aril.

Identification Related to *A. luteola* and *A. varia* but is distinguished by its flat, usually hairless leaflets (exception is ssp. *affinis*). Four subspecies: ssp. *candolleana* (including var. *major*) – consistently 1 pair of pinnae; leaflets 4–11 × 3–7 mm (scattered, Bindoon to near Ravensthorpe).
ssp. *elegans* (synonym *A. pelloiae*) – more than 1 pair of pinnae; gland below insertion of pinnae farthest from stem; leaflets variable, 5–12 × 2–4 mm (near Denman to Many Peaks; Toodyay district).
ssp. *drummondii* – more than 1 pair of pinnae; gland below insertion of pinnae nearest stem; leaflets 3–6 × 1–2 mm (near New Norcia to Collie-Williams area).
ssp. *affinis* – more than 1 pair of pinnae; gland below insertion of pinnae nearest stem; leaflets usually hairy 3–10 × 0·5–1·5 mm (New Norcia to Muchea).

Comments Two subspecies, ssp. *elegans* and ssp. *affinis*, are grown widely in eastern states. The shrub grows best with some shade and shelter in a well-drained position; it is slightly frost tender. Grown from cuttings and seeds.

Acacia drummondii ssp. *elegans*

143 *Acacia cardiophylla*

Group 17

A. Cunn. ex Benth.

Common name	Wyalong Wattle
Meaning of name	Referring to heart-shaped leaflets.
Distribution	On sand or gravel on western slopes and plains of NSW, e.g. in the Wyalong area.
Habit	Soft, fine, ferny-leafed, often bushy shrub to 4 m tall with smooth, grey, slightly ribbed branches, sometimes arched; branchlets rounded, felty hairy. New growth light green and densely hairy.
Foliage	Green, hairy, bipinnate leaves; 12–18 pairs of pinnae with 6–12 pairs of tiny leaflets, densely hairy, oval to heart-shaped, 1–2 mm long; tiny glands, if present, at base of each pair of pinnae.
Flowers	Masses of small, very bright yellow balls 5–6 mm diameter, each of 20–30 flowers, on sparsely hairy short stalks 2–3 mm long, in simple or branched hairy axillary racemes. Flowering July–September.
Pods	Brown, covered with long white hairs, thin-textured 3–7 cm × 5–6 mm with nerve-like margins, raised over and slightly constricted between seeds.
Seeds	Black, oval, 5–6 mm × 2·5 mm × 1·5–2 mm thick, longitudinal in pod; seed-stalk threadlike before thickening into a cap-like aril.
Identification	Tiny, usually heart-shaped leaflets, masses of small bright yellow flower-heads in racemes.
Comments	Very hardy, attractive shrub; tolerates a wide range of conditions from Alice Springs where it is a recommended shrub, to Tas. where it flowers freely. It requires a well-drained position. Grown from cuttings or seeds.

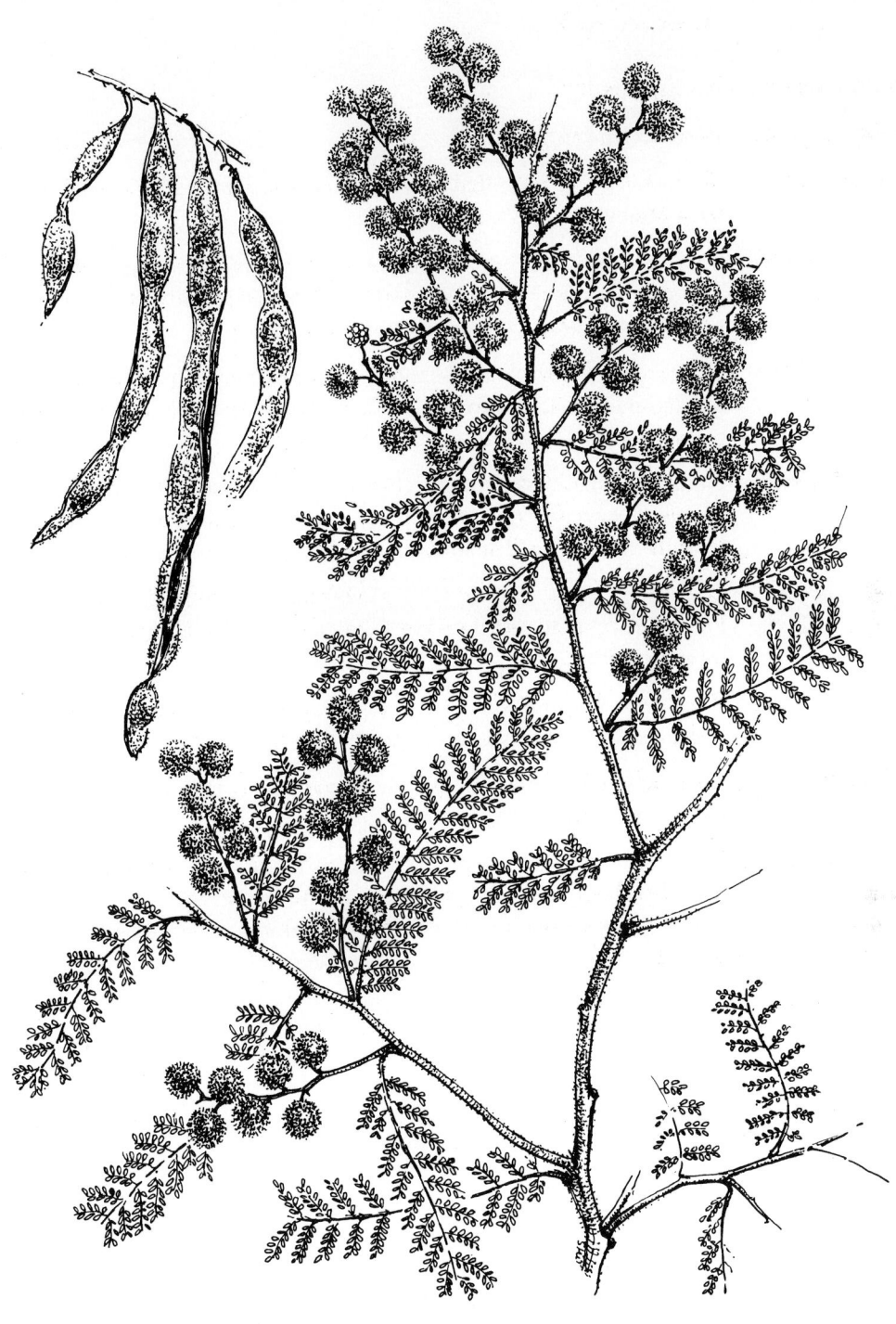

Acacia cardiophylla

144 *Acacia pubescens*
(Vent). R. Br.

Common name Downy Wattle

Meaning of name Downy with soft hairs.

Distribution Confined to a very limited area on gravelly clay ridges and shales of Blue Mountains, NSW, near Bilpin.

Habit Shrub or small tree 3–6 m × 2–3 mm with very fine soft green to grey-green feathery foliage, smooth grey bark and slender branches; branchlets covered with dense erect white hairs.

Foliage Green to grey-green finely bipinnate leaves with 6–12 pairs of pinnae; 6–20 pairs of leaflets, crowded smooth oblong 3–5 mm × 0·75 mm on hairy stems; glands absent.

Flowers Numerous small, fragrant, bright lemon-yellow balls 5–7 mm diameter, each of 15–20 flowers, on smooth or hairy stalks 5 mm long, in slender, usually drooping racemes of 10–20 flowers, up to 6–8 cm long. Flowering July–October.

Pods Red-brown, thin-textured 4–5 (–7) cm × 4–5 mm, slightly contracted between seeds.

Seeds Black, oval 3–4 mm × 2–3 mm, longitudinal in pod; seed-stalk thickened into a whitish, boat-shaped aril.

Identification Long spreading hairs on all branchlets and pinnae stalks; very fine leaflets and small lemon-yellow flower-heads.

Comments This species has been in cultivation for 170 years in Europe; it will tolerate exposed situations and heavy winds, appreciates summer watering but requires good drainage. Grown from seeds. In its natural area it is reported to be a shy seed-setter and of suckering freely after fire. In Tas. it flowers and sets seeds freely and is unaffected by frost. It is listed as 'at risk' by NSW National Parks and Wildlife Service.

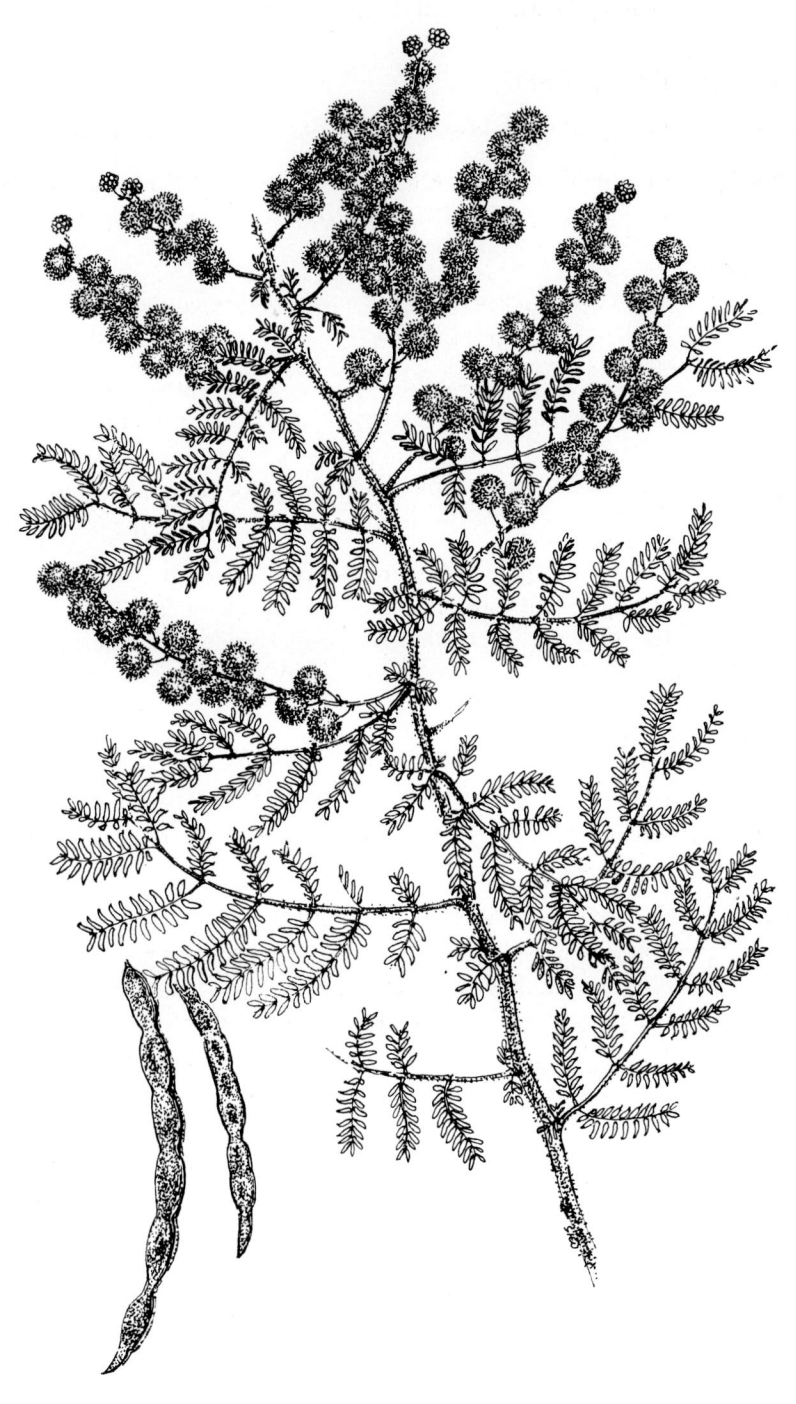

Acacia pubescens

145 *Acacia deanei*

(R. T. Baker) Welch, Coombs et McGlynn

Common name	Deane's Wattle
Meaning of name	Named after H. Deane (1847–1924), a noted amateur botanist.
Distribution	Common and widespread through dry western plains and slopes of NSW, southern Qld (e.g. near Kogan to near Charleville), occasional in Vic. (near Chiltern).
Habit	Ferny-leafed, bushy shrub or small tree 2–7 m tall with grey-brown bark and branches; branchlets angular, slightly ribbed, densely or sparsely covered with appressed hairs. New growth golden yellow.
Foliage	Leaves bipinnate, blue-green; (4–) 6–12 pairs of pinnae on hairy stems; 15–25 (–30) pairs of leaflets, linear, flat, usually clothed with appressed hairs, 2–5 mm long; elongated raised glands usually occur irregularly at or below insertion of each pair of pinnae.
Flowers	Perfumed, lemon balls 6–8 mm diameter, each of 20–25 or more flowers, on shortly hairy stalks 3–5 mm long, in slender racemes of 10–30 flowers, occasionally in panicles. Flowering irregularly throughout the year, but mainly in May–June.
Pods	In clusters, dark brown, straight 5–17 cm × 5–9 mm with nerve-like margins, very little raised over seeds, markedly or slightly constricted between them. Immature pods with a mat of dense, appressed, white hairs.
Seeds	Black, oblong 5–6 × 3–4 mm, longitudinal in pod; seed-stalk thick, short, straight, thickened into a cup-shaped aril.
Identification	Rather like *A. mearnsii*, but differs in being a more hairy, smaller shrub or tree, flowering at a different time and occupying a drier habitat. Ssp. *paucijuga* differs from the type in its more angular branchlets; longer leaflets, usually hairless or sparsely hairy; different flowering time (summer); found in NSW (Grenfell, Temora, Young areas) and Vic. (near Suggan Buggan and Snowy River valley). There are known intermediates between ssp. *deanei* and ssp. *paucijuga*.
Comments	Very hardy, drought and frost resistant species suited to a wide range of soils and conditions. *A. deanei* ssp. *deanei* is widely grown in Qld and is reported to flower almost continuously in Brisbane and Rockhampton. It is useful for windbreaks but was suspected of poisoning sheep and cattle in central NSW some years ago.

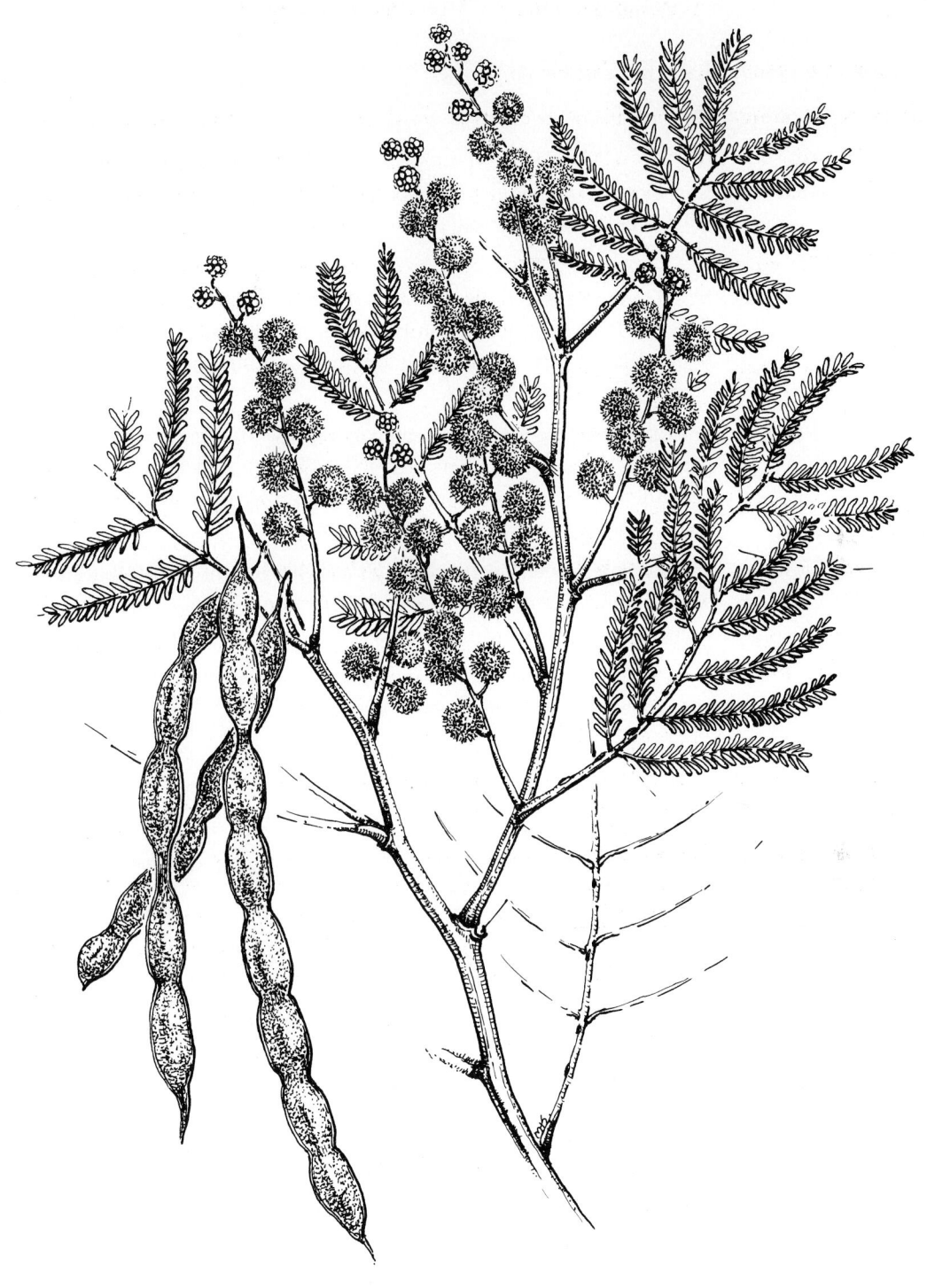

Acacia deanei

146 *Acacia decurrens* (J. Wendl.) Willd.

Common name Early Black or Green Wattle

Meaning of name Running down, referring to wings or ribs running down stems from base of leaves.

Distribution In cooler hills and gullies of southern and central coastal areas and tablelands of NSW. Introduced and now well established in other states.

Habit Small to medium, densely crowned tree 4–12 m tall or more, with smooth, dark green, feathery leaves, dark grey almost black bark; branchlets smooth, angular with longitudinal wings up to 2 mm wide, or ribs running down stems.

Foliage Smooth, dark green, fine, bipinnate leaves; (4–) 6–12 or more pairs of pinnae, 4–7 cm long; leaflets (15–) 30–40 pairs, well-spaced very fine smooth 5–10 (–14) mm long, dark green above, pale beneath; raised gland at or near base of each pair of pinnae.

Flowers Masses of perfumed, small, clear yellow balls, each of 20–30 flowers, on smooth or sparsely hairy stalks 3–6 mm long, in long axillary racemes or panicles, of 6–15 flowers. Flowering July–early September.

Pods Light reddish-brown, flat, almost straight 5–10 cm × 5–8 mm, raised over and somewhat constricted between seeds.

Seeds Black, oval 4–4·5 mm × 2 mm, longitudinal in pod; short seed-stalk thickened into a small aril.

Identification Nearest relatives are *A. mearnsii* and *A. parramattensis*, from which it differs in brighter flowers, earlier flowering and angular ribbed or winged stems.

Comments Fast growing, hardy, frost tolerant tree which has been cultivated for many years and is widely planted as a shade tree or for windbreaks; it is subject to attack by borer, which tends to shorten its life. The timber is reddish-brown, tough and is used commercially as a pulpwood in papermaking. The bark is used in the production of tannin glue, which is used in plywood making.

Acacia decurrens

147 *Acacia muelleriana*

Group 17

Maiden & R. T. Baker

Common name	None known.
Meaning of name	Named for Baron F. von Mueller (1825–1896), renowned early botanist and explorer in Aust.
Distribution	Restricted to shallow, rocky soil in Darling Downs district in forest areas near Chinchilla in Qld, and in areas of north and central western slopes NSW.
Habit	Green, shining, feathery-leafed, rounded shrub or small tree 4–6 m tall, widely spreading up to 6 m with grey round branches; branchlets smooth, green, ribbed, slightly angular at first. New growth bright light green.
Foliage	Dark green bipinnate leaves; 1 or 2, rarely 3 pairs of pinnae; pinnae stalk prominently ribbed or winged on upper edge; 4–6 or more pairs widely spaced, smooth, straight, fine leaflets 10–25 mm × 1–1·5 mm with blunt tip and central nerve; gland at or near base of each pair of pinnae.
Flowers	Numerous, small, cream perfumed balls 5–6 mm diameter, each of 5–12 flowers, on fine smooth stalks 7–8 mm long in axillary racemes of 6–12 flowers, the upper ones forming loose panicles. Flowering September–November.
Pods	Red-brown, thin, smooth, flat 7–18 cm × 5–9 mm with slightly thickened, light coloured margins, either slightly or much constricted between seeds.
Seeds	Black, oval, 5–6 mm × 2–3 mm, longitudinal in pod; seed-stalk folded once and thickened into a club-shaped aril.
Identification	All parts smooth, long, well-spaced, fine leaflets, number of pinnae, winged stem and pale, few-flowered heads.
Comments	Suited to a warm well-drained position in full sun in southern states. It is growing well in NSW and Vic. but is rather slow in Tas. Grown from cuttings and seeds.

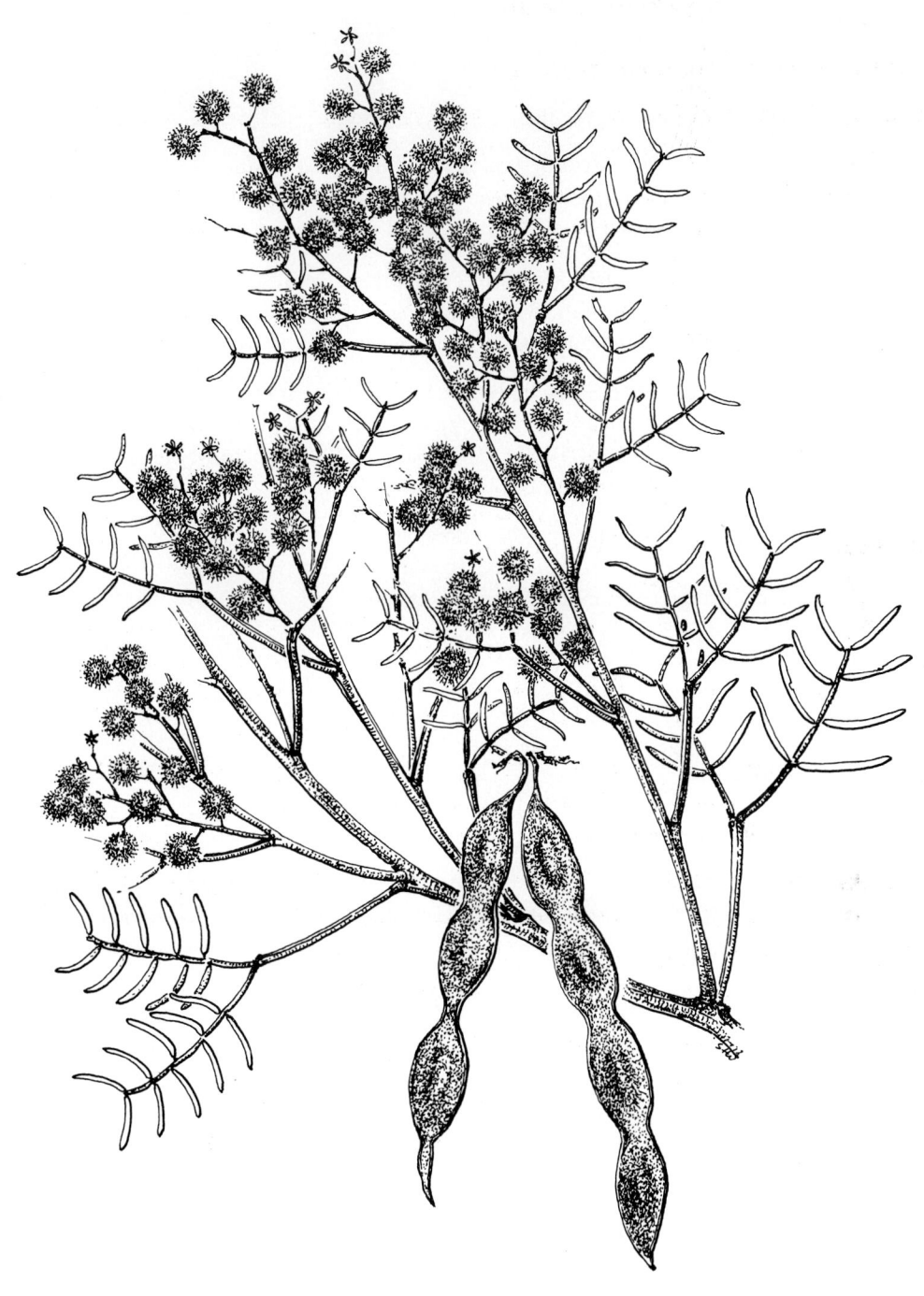

Acacia muellerana

148 *Acacia spectabilis*
A. Cunn. ex Benth.

Common name	Mudgee or Glory Wattle
Meaning of name	Showy, spectacular.
Distribution	Widely distributed on north and central western slopes and plains NSW and into adjacent areas of Qld (Darling Downs district and nearby areas).
Habit	Slender, often spreading, blue-green, ferny-leafed tall shrub or small tree 2–5 m tall, with smooth grey bark and whitish bloom on its often pendulous branchlets. Young growth often purplish.
Foliage	Blue-green, large bipinnate leaves, 2–5 (–7) pairs of pinnae on usually smooth stalks, 10 cm long; 4–8 pairs of leaflets, smooth blunt oblong, faintly veined 5–10 (–13) mm × 2·5–5 mm; gland often small at or below lowest pair of pinnae.
Flowers	Fragrant, large bright yellow balls 8–10 mm diameter, each of 15–20 flowers, on usually smooth stalks in long racemes, the upper ones often in panicles. Flowering July–October.
Pods	Flat, thin-textured, purplish, covered with bloom, 5–11 cm × 10–15 mm with nerve-like margins, raised over and occasionally narrowed between seeds.
Seeds	Black, oblong 5–7 mm × 2–3·5 mm × 2 mm thick, longitudinal in pod; seed-stalk fine, ending in a thickened aril.
Identification	Number of pairs of pinnae, size of leaflets, large racemes of flowers and smooth, whitish trunk.
Comments	Moderately drought and frost resistant. Useful as an ornamental or street tree; widely grown in most states but requires a warm, well-drained position in the south. Grown from cuttings and seeds. A prostrate form is being tried for general cultivation.

Acacia spectabilis

149 *Acacia terminalis*

Salisb.

Synonym	*A. botrycephala* (Vent.) Desf.
Common name	Sunshine Wattle
Meaning of name	Referring to terminal racemes of flowers.
Distribution	Common and widely distributed in coastal areas, woodlands and slopes of NSW, ACT, Vic. and Tas.
Habit	Ferny-leafed, often rounded shrub or small tree 1–6 m tall with smooth mottled bark, spreading branches and angular, ribbed, smooth or hairy branchlets. New growth light bright green, often reddish.
Foliage	Leaves dark green, bipinnate on angular, ribbed stems; 2–6 pairs of pinnae with 10–20 pairs of large, well-spaced leaflets, oblong, smooth, 8–20 mm × 2–5 mm with a central nerve, dark green above, paler below; a large, raised, elongated gland below joint of lowest pair of pinnae, sometimes a few small glands on upper pairs.
Flowers	Large pale to mid-yellow balls 8–10 mm diameter, each of 6–15 flowers, in loose axillary racemes, the upper ones often forming spreading terminal panicles. Flowering March–June, sometimes later.
Pods	Stalked, red-brown, oblong, flat, ± straight 3–10 cm × 10–14 mm with wide margins, occasionally irregularly constricted between seeds. Immature pods often with bright red margins.
Seeds	Large, shiny black, oval 4–5·5 mm × 2·5–3 mm, longitudinal in pod with a fine seed-stalk folded several times and thickened into a boat-shaped aril.
Identification	Large size and number of leaflets and pinnae, flattened pods and time of flowering.
Comments	Very hardy in southern gardens in a variety of soils, full sun or partial shade, but does need some summer moisture. Sometimes short lived and is subject to borer attack. Grown from seeds.

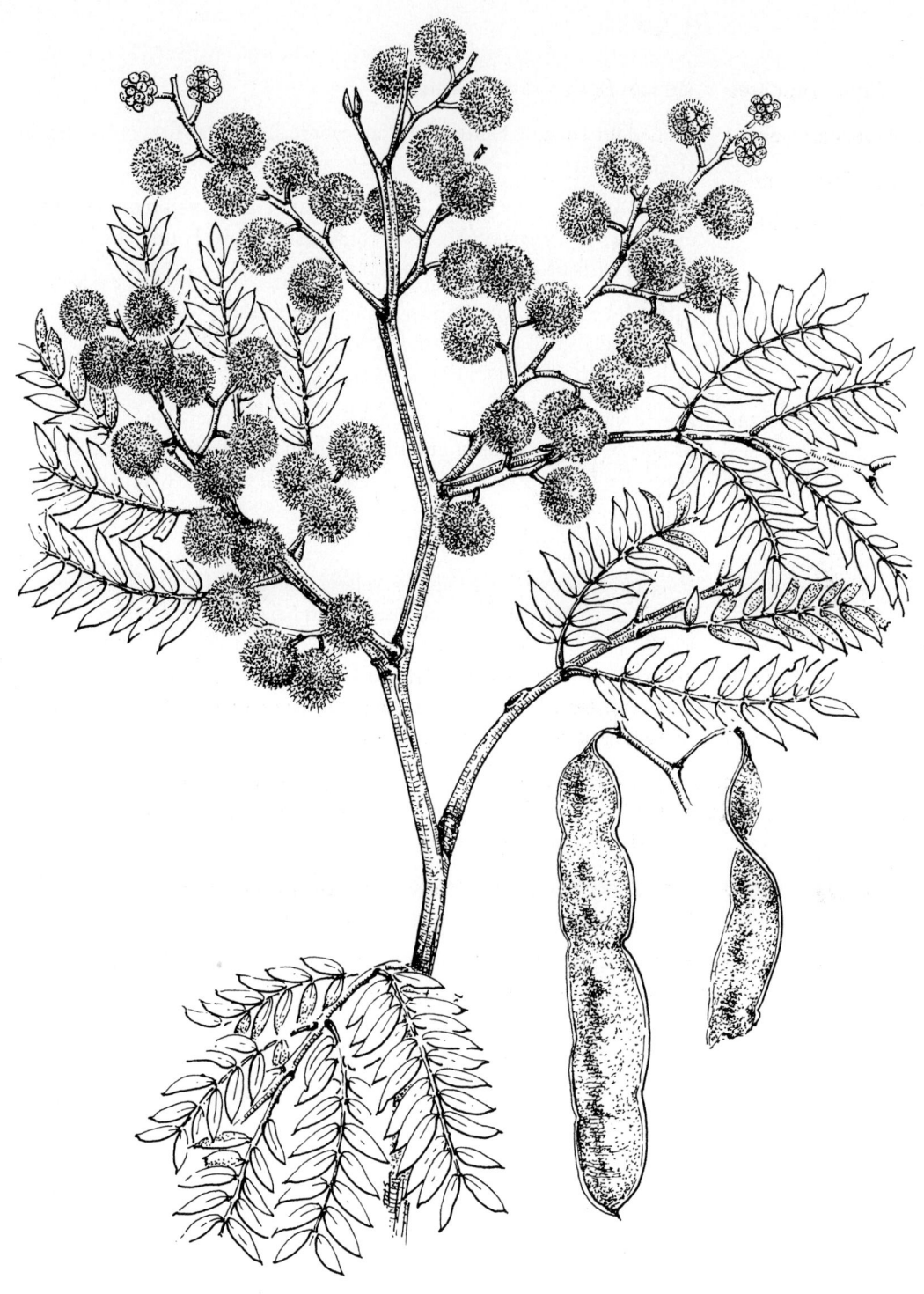

Acacia terminalis

150 *Acacia farnesiana*

Group 17

(L.) Willd.

Common name	Mimosa Bush or Sweet Wattle
Meaning of name	Named for Farnese Gardens in Italy where shrub was first cultivated in 1611.
Distribution	Widespread in sandy loams, black soils, often along watercourses of the more northern parts of Australia from east to west. Found also in tropical parts of America, Asia and Africa.
Habit	A much-branched, very thorny, spreading, feathery-leafed shrub or occasionally small tree 1–7 m tall with grey-brown bark, smooth or scaly, lined branchlets, often zig-zagging.
Foliage	Leaves bipinnate; 2–4 (–7) pairs of pinnae on sparsely hairy stems; usually 8–20 pairs of leaflets smooth green or blue-green linear with blunt tips 3–6 (–9) × 1–2 mm; a large gland at or between each pair of pinnae; two very sharp-pointed spines 2–30 mm long or longer at base of pinnae stalks, sometimes absent.
Flowers	Very large, dense, bright orange-yellow fragrant balls 15–20 mm diameter, each of 50 or more flowers, on hairy or hoary stalks 8–20 (–25) mm long, sometimes singly but usually two or three grouped together in axils, with small bracts close under the flowers. Flowering mainly June–September; rain or soil moisture influences flowering time.
Pods	Black or dark brown, smooth, thick, irregularly rounded, often lined, straight or curved, 4–8 cm × 8–12 mm. Pods often remain unopened on bushes for a long time.
Seeds	Large, grey-black, 6–7 mm × 4–5 mm, packed in a white pithy substance; transverse or oblique in pod; short, straight seed-stalk.
Identification	Large, fragrant, orange-yellow flowers, very long spines, black round pods packed with pith.
Comments	Grown from seeds; suitable as a low shelter shrub – birds use it as a safe nesting site. It is reported that pods are eaten by sheep. The highly perfumed flowers are used commercially for the production of an oil with the odour of violets, which is used in the scent industry. Its timber is close-grained, heavy, and has been recorded as taking a good polish. The shrub will not tolerate severe southern frosts.

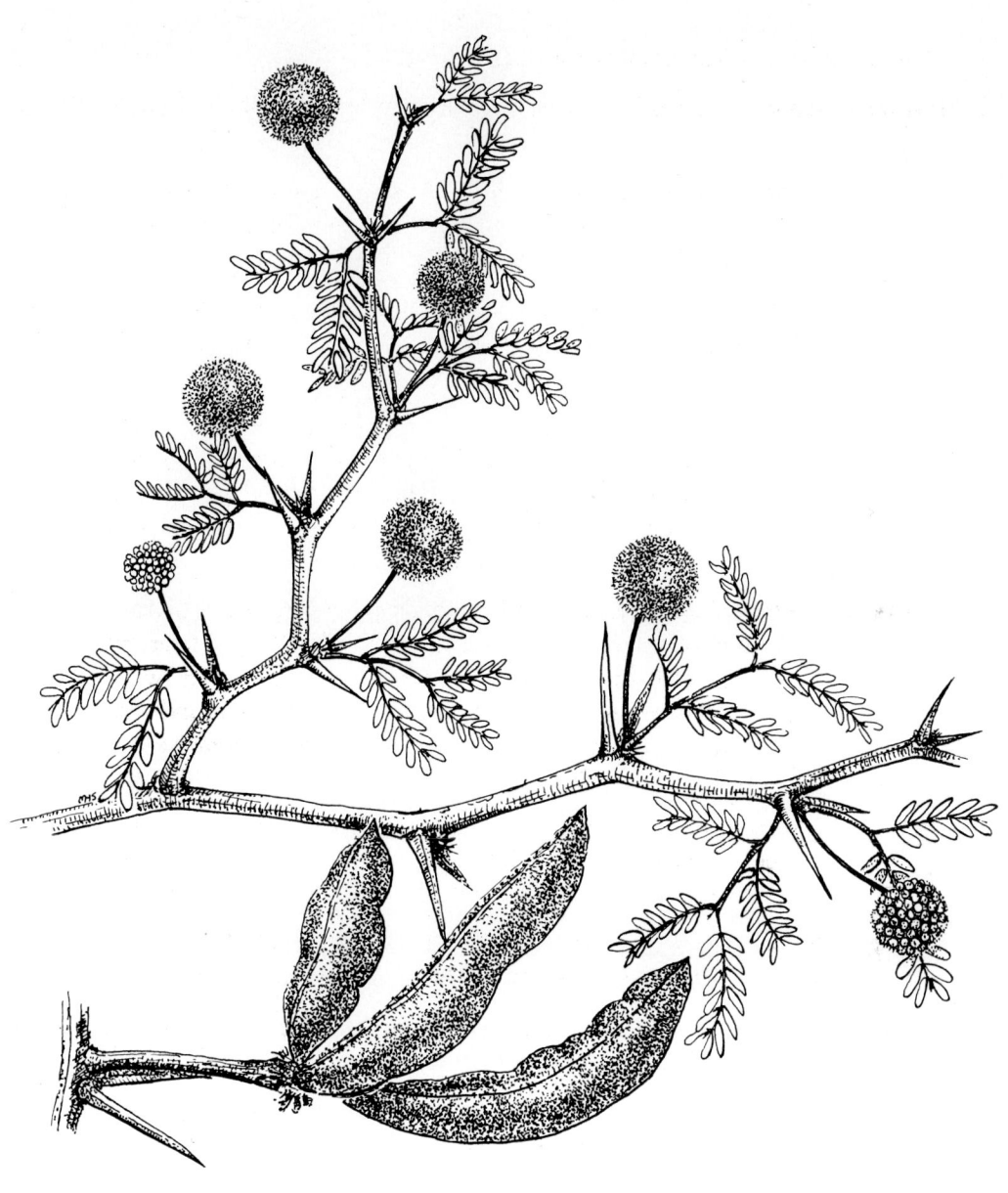

Acacia farnesiana

Naming Plants

It is customary for a plant's botanical name to be followed by that of the person who first described and published its details. Well-known names are often abbreviated.

In some instances two names are listed, e.g. '(Andr.) Willd.'. The name of the original author is bracketed and followed by the name of the person who placed the plant in its current category.

Where two names are linked by 'ex', e.g. 'A. Cunn. ex Benth.', it indicates that Cunningham suggested the plant name and Bentham described and published it.

Where 'f.' follows a name, it indicates the son of the named well-known botanical author.

C. Andrews	Cecil R. P. Andrews, 1870–1951
(Andr.) Willd.	(Henry C. Andrews, ?–1830), Karl. L. Willdenow, 1765–1812
F. M. Bail.	Frederick M. Bailey, 1827–1915
R. T. Baker	Richard T. Baker, 1855–1941
(R. T. Baker) Welch, Coombs et McGlynn	(Richard T. Baker, 1855–1941), –. Welch, 1895–1942
Benth.	George Bentham, 1800–1884
J. M. Black	John M. Black, 1855–1951
R. Br.	Robert Brown, 1773–1858
(Burm.f.) Hort ex Hoffmannsegg	(Nicolaus L. Burman, 1734–1793), Hort, means of garden origin, Johann C. G. von Hoffmannsegg, 1766–1849
A. Cunn. ex Benth.	Allan Cunningham, 1791–1839, George Bentham, 1800–1884
A. Cunn. ex G. Don	Allan Cunningham, 1791–1839, George Don, 1798–1856
A. Cunn. ex Hook.	Allan Cunningham, 1791–1839, William J. Hooker, 1785–1865
DC	Augustin P. de Candolle, 1778–1841
Gard.	Charles A. Gardner, 1896–1970
Henslow	John Henslow, 1796–1861
Ker	John Bellenden Ker, 1765–1842
(Labill.) H. Wendl.	(Jacques J. H. de Labillardiere, 1755–1834), Hermann Wendland, 1823–1903
(L.) Willd.	(Carl von Linné (Linnaeus), 1707–1788), Karl L. Willdenow, 1765–1812

Lehm.	Johanne G. C. Lehmann, 1792–1860
Lindl.	John Lindley, 1799–1865
Macbride	James F. Macbride, present
Maiden	Joseph H. Maiden, 1859–1925
Maiden & Blakely	Joseph H. Maiden, 1859–1925, William F. Blakely, 1875–1941
(Maiden) Maiden & Blakely	Joseph Maiden, 1859–1925, William F. Blakely, 1875–1941
Maslin	Bruce Maslin, present
Meisn.	Carl. F. Meisner, 1800–1874
F. Muell.	Baron Ferdinand von Mueller, 1825–1896
F. Muell. ex Benth.	Baron Ferdinand von Mueller, 1825–1896, George Bentham, 1800–1884
Morris	Dennis L. Morris, present
Pedley	Les Pedley, present
E. Pritzel	Ernst Pritzel, 1875–?
F. M. Reader	?1853–1911
Reichb.	Heinrich G. L. Reichenbach, 1793–1879
Salisb. (Court)	Richard A. Salisbury, 1761–1829, (Arthur B. Court, present)
Sieber ex DC.	Franz W. Sieber, 1789–1844, Augustin P. de Candolle, 1778–1841
(Sm.) Willd.	(Sir James E. Smith, 1759–1828), Karl L. Willdenow, 1765–1812
Sweet ex Lindl.	Robert Sweet, 1783–1835, John Lindley, 1799–1865
Tindale	Dr. Mary D. Tindale, present
(Tindale) Pedley	(Dr. Mary D. Tindale, present), Les Pedley, present
(Vent.) R. Br.	(Etienne P. Ventenat, 1757–1808), Robert Brown, 1773–1858
(Vent.) Willd.	(Etienne P. Ventenat, 1757–1808), Karl L. Willdenow, 1765–1812
Wendl.	Johann C. Wendland, 1755–1828
(J. Wendl.) Willd.	(Johann C. Wendland, 1755–1828), Karl L. Willdenow, 1765–1812

Bibliography

Books

Anderson, R. H.   *The Trees of New South Wales*. Govt Printer, Sydney, 1968.

Armitage, I.   *Acacias of New South Wales*. NSW Region, Society for Growing Australian Plants, Sydney, 1978.

Australian Systematic Botany Society.   *Flora of Central Australia*. Reed, Sydney, 1981.

Bailey, F. M.   *The Queensland Flora*. Diddams & Co., Brisbane, 1900.

Beadle, N. C. W.,   *Student's Flora of North Eastern New South Wales*. Part 3. University of New England, Armidale, 1976.

Beadle, N. C. W., Evans, O. D., Carolin, R. C. and Tindale, M. D.   *Flora of the Sydney Region*. Reed, Sydney, 1972.

Beard, J. S.   *Descriptive Catalogue of West Australian Plants*. Soceity for Growing Australian Plants, Sydney, 1970.

Bentham, G.   *Flora Australiensis*. Lovell Reeve & Co., London. Reprint 1967.

Black, J. M.   *Flora of South Australia*. Part 2. Govt Printer, Adelaide, 1963.

Boomsma, C. D.   *Native Trees of South Australia*. Woods & Forests Department, Adelaide, 1972.

Burbidge, N. T. and Gray, M.   *Flora of the Australian Capital Territory*. Australian National University Press, Canberra, 1970.

Canberra Botanic Gardens.   *Growing Native Plants*. Department of the Capital Territory. Vols 1–8, 1971–78.

Chandler, B.   *Two Hundred Wattles for Gardens*. David Stead Memorial Wildlife Research Foundation of Australia, Sydney, 1975.

Cochrane, G. R., Fuhrer, B. A., Rotherham, E. R., Simmons, J. & M. and Willis, J. H.   *Flowers and Plants of Victoria and Tasmania*. Reed, Sydney, 1980.

Costermans, L.   *Native Trees and Shrubs of South-eastern Australia*. Rigby, Melbourne, 1983.

Cunningham, G. M. *et al.*   *Plants of Western New South Wales*. Soil Conservation Service of New South Wales, Sydney, 1981.

Curtis, W. M. and Morris, D. I.   *The Student's Flora of Tasmania*. Part 1. 2nd edition. Govt Printer, Hobart, 1975.

Curtis, W. M. and Stones, M.   *Endemic Flora of Tasmania*. Part 2. Ariel Press, London, 1969.

Debenham, C.   *The Language of Botany*. Society for Growing Australian Plants, Sydney.

Diels, L. and Pritzel, E.   *Fragmenta Phytographiae Australiae Occidentalis*. Bot. Jhb. Leipzig, 1904–5.

Don, G.   *General History of Dichlamydeous Plants*. Vol. 2, 1832.

Eichler, H.   *Supplement to J. M. Black's Flora of South Australia*. 2nd edition. Govt Printer, Adelaide, 1965.

Elliot, G.   *Australian Plants for Small Gardens and Containers*. Hyland House, Melbourne, 1979.

Elliot, W. R. and Jones, D. L.   *Encyclopaedia of Australian Plants*. Vol. 2. Lothian, Melbourne, 1982.

Erickson, R., George, A. S., Marchant, N. G. and Morcombe, M. K. *Flowers and Plants of Western Australia*. Reed, Sydney, 1973.

Ewart, A. J. and Davies, O. B. *The Flora of the Northern Territory*. McCarron, Bird & Co., Melbourne, 1917.

Fairhall, A. R. *West Australian Native Plants in Cultivation*. Permagon Press, Sydney, 1970.

Galbraith, J. *Field Guide to the Wildflowers of South-east Australia*. Collins, Sydney, 1977.

Hall, N. *et al.* *The Use of Trees and Shrubs in the Dry Country of Australia*. Department of National Development, Canberra, 1972.

Harmer, J. *Northern Australian Plants*. Society for Growing Australian Plants, Sydney.

Harris, T. Y. *Alpine Plants of Australia*. Angus & Robertson, Melbourne, 1970. *Gardening with Australian Plants: Shrubs*. Nelson, Melbourne, 1977.

Holliday, I. and Hill, R. *A Field Guide to Australian Trees*. Rigby, Adelaide, 1969.

Horton, H. *Around Mt. Isa*. University of Queensland Press, Brisbane, 1976.

Jessop, J. P. and Toelken, H. R. (editors) *Flora of South Australia*. Part 2. State Herbarium of South Australia, South Australian Government Printing Division, Adelaide, 1986.

Marchant, N. G. *et al.* *Flora of the Perth Region*. Part 1. Western Australian Herbarium, Department of Agriculture, Western Australia, 1987.

Mueller, F. von *Fragmenta Phytographiae Australiae*. Iconography of Australian Species of Acacia. 1887–88.

Newbey, K. *West Australian Plants for Horticulture*. Parts 1 and 2. Society for Growing Australian Plants, Sydney.

Rogers, F. J. C. *Growing Australian Native Plants*. Nelson, Melbourne, 1971. *Growing More Australian Native Plants*. Nelson, Melbourne, 1975. *Field Guide to Victorian Wattles*. 2nd edition. The author, 1978.

Rotherham, E. R., Briggs, B., Blaxell, D. F. and Carolin, R. C. *Flowers and Plants of New South Wales and Southern Queensland*. Reed, Sydney, 1975.

Sharr, F. A. *West Australian plant names and their meanings*. University of Western Australia Press, Nedlands, 1978.

Simpfendorfer, K. J. *An introduction to Trees for South-eastern Australia*. Inkata Press, Melbourne, 1975.

Society for Growing Australian Plants. *A Horticultural Guide to Australian Plants*. A series, parts 1–6, Brisbane.

Specht, R. L. and Mountford, C. P. (editors) *Records of the American–Australian Arnhem Land Scientific Expedition*. Vol. 3, 1958.

Stanley, T. D. and Ross, E. M. *Flora of South-eastern Queensland*. Vol. 1. Queensland Department of Primary Industries, Brisbane, 1983.

Turnbull, J. W. (editor) *Multipurpose Australian Trees and Shrubs*. Australian Centre for International Agricultural Research, Canberra, 1986.

Whibley, D. J. E. *Acacias of South Australia*. Govt Printer, Adelaide, 1980.

Willis, J. H. *A Handbook of Plants in Victoria*. Vol. 2. Melbourne University Press, 1972.

Wrigley, J. W. and Fagg, M. *Australian Native Plants*. Collins, Sydney, 1979.

Zimmer, G. F. *Botanical Names and Terms*. Routledge & Kegan Paul, London, 1963.

Journals

Bibliotheca Botanica, Stuttgart, 1886–1926.
Chemist and Druggist of Australasia, 1882–87.
Journal of Botany, British and Foreign, London, 1912–26.
Journal of the Linnean Society (Botany), London, 1899–1920.
Journal of the WA Natural History Society, 1904–27.
Journals issued periodically by Australian Herbaria: Contributions from NSW National Herbarium, now issued as *Telopea*; Contributions from Qld Herbarium, now issued as *Austrobaileya*; *Nuytsia* from WA Herbarium; *Muelleria* from Victoria; the Journal of the Adelaide Botanic Garden from South Australia; and the *Herbarium Australiense* from CSIRO, Canberra.
Journals of the Royal Societies of
  New South Wales, 1915–27.
  Queensland, 1918–64.
  South Australia, 1958–59.
  Victoria, 1888.
  Western Australia, 1923 et seq.
Proceedings of the Linnean Society of New South Wales, 1895–97; 1960–66.
Queensland Agricultural Journals.
Scottish Botanical Review, 1912.
Southern Science Record, 1882 et seq.
Victorian Naturalist, 1893 et seq.

Index

Page numbers in italics refer to line illustrations. Numbers prefixed by the
letters Pl. refer to colour plates.